Signals and Communication Technology

This series is devoted to fundamentals and applications of modern methods of signal processing and cutting-edge communication technologies. The main topics are information and signal theory, acoustical signal processing, image processing and multimedia systems, mobile and wireless communications, and computer and communication networks. Volumes in the series address researchers in academia and industrial R&D departments. The series is application-oriented. The level of presentation of each individual volume, however, depends on the subject and can range from practical to scientific.

Indexing: All books in "Signals and Communication Technology" are indexed by Scopus and zbMATH

For general information about this book series, comments or suggestions, please contact Mary James at mary.james@springer.com or Ramesh Nath Premnath at ramesh.premnath@springer.com.

Saad El Jaouhari • Nour El Madhoun
Editors

Blockchain-based Web of Things Security: Advances, Challenges, and Solutions

Springer

Editors
Saad El Jaouhari
Isep
Paris, France

Nour El Madhoun
Isep
Paris, France

ISSN 1860-4862 ISSN 1860-4870 (electronic)
Signals and Communication Technology
ISBN 978-3-032-07670-0 ISBN 978-3-032-07671-7 (eBook)
https://doi.org/10.1007/978-3-032-07671-7

This Springer imprint is published by the registered company Springer Nature Switzerland AG
The registered company address is: Gewerbestrasse 11, 6330 Cham, Switzerland

Preface

The Internet of Things (IoT) has emerged as one of the most transformative technological advancements of the twenty-first century, fundamentally changing how we interact with the digital and physical worlds. At its core, IoT refers to the seamless interconnection of everyday objects, ranging from household appliances and wearable devices to industrial machinery and smart city infrastructures, capable of exchanging, and processing data over the Internet. This interconnected ecosystem allows devices to operate autonomously, make intelligent decisions, and enhance efficiency across various domains, including healthcare, transportation, manufacturing, and agriculture. The promise of IoT lies in its ability to create a more connected and intelligent world, where data-driven insights optimize operations and improve quality of life. However, as the number of IoT devices grows exponentially, challenges such as cybersecurity threats, interoperability issues, and data privacy concerns continue to shape its evolution. Despite these challenges, the relentless advancement of technologies such as artificial intelligence, blockchain, and edge computing is further refining IoT, pushing it toward more secure, scalable, and intelligent applications. As IoT continues to expand, its integration with other emerging technologies is set to redefine the way people, devices, and environments interact, paving the way for a future where everything is seamlessly connected.

The transition from the Internet of Things (IoT) to the Web of Things (WoT) marks a shift from simply connecting physical devices to enabling seamless interaction and interoperability using web technologies. While IoT focuses on linking devices via the Internet, it often results in fragmented ecosystems due to proprietary protocols and lack of standardization. WoT addresses this by integrating IoT devices into the World Wide Web, using standard web protocols such as HTTP, WebSockets, and RESTful APIs to ensure seamless communication across diverse platforms. This approach enhances interoperability, simplifies development, and allows IoT applications to be more accessible and scalable. By leveraging the Web's open architecture, WoT transforms IoT from a collection of isolated smart devices into a fully integrated and interactive digital ecosystem.

The World Wide Web has evolved significantly since its inception, progressing from a simple network for document sharing to a vast ecosystem that connects billions of physical devices—a concept known as the Web of Things (WoT). The WoT aims to integrate everyday objects—including wearables, home appliances, vehicles, and industrial equipment—into the Internet, enabling seamless interaction and communication among them. This integration allows computers and systems to perceive, interpret, and respond to the physical environment more effectively.

The global number of IoT devices is expected to reach 40.6 billion by 2034. This rapid expansion underscores the increasing prevalence of IoT devices in various sectors, including smart homes, healthcare, transportation, and industrial automation. The WoT leverages existing web standards to ensure interoperability across diverse IoT platforms and application domains, addressing the fragmentation often seen in the IoT landscape.

The increasing adoption of WoT brings unprecedented opportunities but also introduces significant challenges, particularly concerning security, trust, and privacy. As the number of Internet-connected devices continues to grow exponentially, traditional security models struggle to keep pace. Blockchain technology has emerged as a promising solution to enhance the security and trustworthiness of WoT ecosystems. By leveraging decentralization, transparency, and cryptographic security, blockchain-based approaches offer novel solutions for securing communications, managing identities, and ensuring data integrity in WoT environments.

This book provides a comprehensive exploration of the intersection of blockchain technology and the Web of Things. It consolidates the latest research, methodologies, and applications aimed at addressing the security challenges associated with WoT. The book is structured into seven chapters, each offering in-depth insights into various aspects of blockchain-enabled WoT security:

- **Chapter 1**: From the Internet of Things to the Web of Things
- **Chapter 2**: Introduction to Blockchain Technology
- **Chapter 3**: Blockchain-Based Approaches for Securing Web of Things (WoT)
- **Chapter 4**: A Smart Contract-Based Blockchain Solution for Key Revocation in IoT Networks
- **Chapter 5**: Security of Intelligent Transport Systems Using Blockchain: A Review of the Literature
- **Chapter 6**: A Blockchain Solution for Reliable and Secure Distributed SSO Application
- **Chapter 7**: Building Trust in the Web of Things: Decentralized Identity Management with Blockchain

By compiling cutting-edge research and practical solutions, this book serves as a valuable reference for researchers, practitioners, and policymakers seeking to navigate the evolving landscape of WoT security. It provides insights into the fundamental principles, key challenges, and innovative blockchain-based solutions that contribute to building a secure and resilient WoT ecosystem.

Paris, France

Saad El Jaouhari
Nour El Madhoun

Acknowledgments We would like to express our sincere and heartfelt thanks to all those who contributed to the success of this collective work. Each chapter is the result of a rich collaboration, and we are truly grateful for the time, effort, and expertise generously shared throughout the process. We are especially thankful to the following colleagues for their valuable contributions to the scientific quality of the book and for their careful reviewing and insightful comments: Nouredine Tamani (Isep, France), Abdul Qadir Khan (Isep, France), Pengwenlong Gu (CNAM, France), Chougdali Khalid (ENSA, Morocco), Khaoula Boukir (Ibn Tofail University, Morocco), Badis Hammi (Telecom SudParis, France), Daniel Alejandro Maldonado Ruiz (Escuela Politécnica Nacional, Ecuador), Ioanna Dionysiou (University of Nicosia, Cyprus), Jenny Torres (Escuela Politécnica Nacional, Ecuador), Michal Kit (Telecom SudParis, France), Montassar Bellah Naghmouchi (Telecom SudParis, France). Their rigor, generosity, and collegial spirit were instrumental in ensuring the quality and coherence of this publication.

We would also like to extend our warmest thanks to the Springer team, whose continuous support and professionalism were crucial throughout the editorial process. We are deeply grateful to Brian Halm, Kaitlyn Bonomo, Mary James, Olivia Ramya Chitranjan, and Pradheepa Vijay for their availability, patience, and commitment. Their guidance and dedication made this experience smooth and enjoyable from start to finish.

Contents

About the Editors

Dr. Saad El Jaouhari is currently an Associate Professor at Isep—Engineering School in Paris. He is a member of the Electronics and Communication Systems (ECoS) team within the LISITE laboratory (Laboratoire d'Informatique, Signal et Image, Électronique et Télécommunications). He also serves as the Head of the Cybersecurity Lab, and leads both the engineering cycle and the master's program in Wireless Telecommunication and IoT Systems. He obtained his master's degree in Cryptology and Computer Security from Bordeaux University in 2015. In December 2018, he earned his Ph.D. in Computer Science from IMT Atlantique, France, where his research focused on the cybersecurity of smart cities and Internet of Things/Web of Things services. Following his Ph.D., he joined IRT b<>com as a postdoctoral researcher in 2019, contributing to the CONDOR-SET project, which aimed to stream operating room videos enriched with metadata. In 2020, he continued his postdoctoral work at Orange Labs, focusing on over-the-air firmware updates for IoT devices. In 2024, he held a temporary position as Director of Research and Innovation at Adservio, where he led the research team, oversaw project planning, and managed the technology intelligence process. His research interests include cybersecurity, the Internet and Web of Things (IoT/WoT), and Software-Defined Networking (SDN). His current work centers on enhancing the security of IoT networks, with a particular emphasis on anomaly detection. This involves investigating semantical and statistical techniques to enable comprehensive analysis of IoT data and network behavior.

Nour El Madhoun is currently Associate Professor of Computer Science, Cybersecurity and Blockchain and Head of the engineering cycle Digital Security and Networks at Isep—Engineering School in Paris, a position she has held since May 2022. She is also an Associate Researcher at Sorbonne Université/LIP6, where she was part of the PHARE team from 2020 to 2022 and has been a member of the NPA team since November 2022. In addition, she is affiliated as an Associate Researcher with Université Paris-Saclay/LISN in the ROCS team.

She received her master's degree in Networks/-Computer Science from Sorbonne Université/Télécom ParisTech in 2014 and her Ph.D. degree in Cybersecurity/Computer Science from Sorbonne Université in 2018. At Sorbonne Université, she served as an ATER in 2017. In 2018, she gained industry experience as a post-doctoral researcher at Orange Labs, working on blockchain and smart contract technologies.

From 2019 to 2020, she joined Isep—Engineering School in Paris as an Associate Professor of Cybersecurity and Blockchain, while also overseeing the engineering cycle Digital Security and Networks. From 2020 to 2022, she served as an Associate Professor of Cybersecurity and Blockchain at EPITA—Engineering School in Paris.

Her current research focuses on network security, cryptographic protocols, EMV payment systems, NFC technology, blockchain and smart contract technologies, and Graph Neural Networks (GNNs) for cybersecurity.

She has co-edited the following four Springer books: *Emerging Trends in Cybersecurity Applications* (with Kevin Daimi, Abeer Alsadoon, Cathryn Peoples), *Principles and Practice of Blockchains* (with Kevin Daimi, Ioanna Dionysiou), *Building Cybersecurity Applications with Blockchain and Smart Contracts* (with Ioanna Dionysiou, Emmanuel Bertin), *Blockchain and Smart Contract Technologies for Innovative Applications* (with Ioanna Dionysiou, Emmanuel Bertin).

From the Internet of Things to the Web of Things

Saad El Jaouhari

1 Introduction

To provide a solid foundation for this book on Blockchain-Based Web of Things Security, it is essential to first introduce and clearly define the concept of the Web of Things (WoT). This chapter primarily aims to present a comprehensive overview of the WoT, beginning with a detailed examination of its abstract architecture and exploring two representative use cases that highlight its practical applications. Following this, the chapter analyzes the principal security threats related to the WoT, emphasizing the unique risks that arise from the convergence of web technologies and IoT devices (or Things). Finally, it reviews notable secure architectural solutions proposed in the literature, setting the stage for understanding how blockchain technologies can effectively enhance WoT security. The chapter is structured as follows: In Sect. 2, we first present the notion of IoT in a nutshell. Then, in Sect. 3, we introduce the notion of the WoT, its abstract architecture and its core components, and some illustrative use cases. In Sect. 4, the main security threats are listed, and finally, before concluding in Sect. 6, we present in Sect. 5.2 two different architectures proposed in the literature to secure the WoT.

S. El Jaouhari (✉)
Isep, LISITE, Issy-les-Moulineaux, France
e-mail: saad.el-jaouhari@isep.fr

S. El Jaouhari, N. El Madhoun (eds.), *Blockchain-based Web of Things Security: Advances, Challenges, and Solutions*, Signals and Communication Technology,
https://doi.org/10.1007/978-3-032-07671-7_1

2 The Internet of Things (IoT) in a Nutshell

In recent years, the IoT has emerged as a transformative paradigm with far-reaching applications across virtually all sectors, including healthcare, transportation, manufacturing, and urban infrastructure. Although the term "IoT" has gained widespread adoption, there remains no universally agreed-upon definition, as the concept continues to evolve from both technological and societal perspectives. Broadly defined, the IoT refers to a dynamic and distributed network of interconnected physical and virtual "Things," ranging from sensors and actuators to everyday objects and even people, which are capable of sensing, processing, exchanging, and acting on data. The integration of these connected entities into digital environments is expected to profoundly reshape everyday life, especially in the context of smart cities, where urban systems and services become increasingly intelligent, efficient, and responsive. These "Things" typically consist of resource-constrained embedded systems equipped with communication, sensing, and basic processing capabilities. Despite their limitations, such as low computational power, limited memory and storage, restricted communication range, and finite energy supplies (often due to battery constraints), they are able to interact both with users and with other devices. These Smart Objects (SOs), as they will be referred to throughout this thesis, can function in Machine-to-Human (M2H) systems by providing data or receiving commands from users, or in Machine-to-Machine (M2M) configurations, where they autonomously exchange information with other devices. This ubiquitous connectivity enables a broad range of innovative use cases, from low-cost environmental monitoring to the deployment of complex, automated systems within the emerging "Smart City" landscape.

The application domains of IoT technologies are both numerous and diverse. Among the most prominent areas is healthcare, where an increasing number of well-being and medical devices are being introduced to the market. These devices can monitor vital signs and track patient conditions, whether in hospitals or at home. In smart industry, the shift toward Industry 4.0 emphasizes the development of intelligent production systems and interconnected manufacturing environments, where IoT plays a central role. In smart homes and buildings, connected devices such as intelligent thermostats, lighting, and security systems are gaining widespread adoption. Similarly, smart energy systems utilize smart plugs and meters to monitor and optimize the consumption of electricity, gas, water, and heat. Finally, Intelligent Transportation Systems (ITS) represent a significant area of innovation, with smart and autonomous vehicles capable of sensing their environment and adapting in real time to enhance safety and user experience.

As previously mentioned, a key feature of smart objects is their ability to communicate both with each other and with humans. To enable this interaction, a variety of wireless communication technologies and protocols have emerged. In the context of IoT, these can generally be divided into two main categories: short-range technologies—such as 6LoWPAN, Zigbee, Bluetooth, Z-Wave, NFC, and RFID—and long-range technologies—such as SigFox, LoRa, and cellular networks, which are typically classified under Low-Power Wide-Area Networks (LPWAN). In this thesis, only short-range communication protocols have been implemented as a proof of concept. Nonetheless, the proposed concepts are equally applicable to long-range protocols. However, further analysis would be required, particularly in relation to security and privacy considerations.

Several architectural frameworks have been proposed to support the development of IoT systems, including oneM2M [11], IoT-A [9], and initiatives from the Alliance for Internet of Things Innovation (AIOTI) [3], among others. Among these, oneM2M is one of the most mature and widely adopted standards, with publicly available specifications and several open-source implementations. However, a noted limitation of the oneM2M architecture is its primary focus on Machine-to-Machine (M2M) communication, with less emphasis on direct end-user (human) involvement. In addition to these, the Open Connectivity Foundation (OCF) proposes a comprehensive framework focused on secure and interoperable IoT communications, particularly at the device level [2]. Similarly, the Lightweight Machine-to-Machine (LwM2M) protocol [1], developed by the Open Mobile Alliance (OMA), offers an efficient architecture for managing constrained devices, especially in environments where bandwidth, memory, and power are limited.

Despite these advancements, many of these architectures have been developed independently and lack native interoperability with one another. This leads to the creation of isolated ecosystems, or silos, which pose challenges to seamless integration, cross-platform communication, and broad-scale deployment. The novelty of the IoT does not lie solely in the functional capabilities of individual smart objects, since many embedded systems are already connected to the Internet, but rather in the sheer scale of interconnected devices anticipated in the coming years. The prospect of deploying billions or even trillions of smart objects introduces a range of unprecedented technical and societal challenges that go far beyond traditional embedded systems.

According to market research by IoT Analytics reports approximately 16.6 billion connected IoT devices by the end of 2023, with forecasts reaching 18.8 billion by the end of 2024 and $\sim$40 billion by 2030 [10]. This explosive growth underscores the urgency of addressing a variety of critical challenges, including but not limited to the following, and as shown in Fig. 1:

- Security and Privacy: ensuring secure communication, data integrity, and protection of user information in highly distributed environments
- Dynamic Discovery and Management: allowing devices to autonomously discover, register, and interact within heterogeneous networks
- Interoperability and Fragmentation: overcoming incompatibilities at both the architectural level (as discussed earlier) and the communication protocol level, where many technologies are not designed to work together
- Scalability and Data Handling: managing the vast amounts of data generated by diverse and resource-constrained devices

Fig. 1 Main Internet of Things challenges

Table 1 Heterogeneous IoT ecosystems

Device	Manufacturer	Protocol	Ecosystem
Nest Thermostat	Google	Google Protocol, Wi-Fi	Google Home
Echo Dot	Amazon	Alexa API, HTTP	Amazon Alexa
Hue Bulb	Philips	Zigbee, Bluetooth	Philips Hue App
SmartThings Hub	Samsung	Zigbee, Z-Wave, Wi-Fi	SmartThings
Arlo Camera	Arlo (Netgear)	Arlo Protocol, Wi-Fi	Arlo Secure App

– Energy Efficiency: A significant number of IoT devices operate on limited battery power, often in remote, hard-to-access, or embedded environments where frequent recharging or battery replacement is difficult.

For instance, as shown in Table 1, different IoT devices utilize distinct communication protocols, highlighting the interoperability challenges in WoT systems.

These challenges are fundamental to the successful and sustainable deployment of IoT systems at global scale.

3 From IoT to the Web of Things (WoT)

To overcome the siloed nature of the IoT and unify its fragmented landscape of specifications, architectures, and communication protocols, the Web of Things (WoT) was introduced through the efforts of the W3C. It can be viewed as a specialization or extension of the IoT, which aims to simplify the development and integration of IoT applications by leveraging web standards. On the one hand, WoT abstracts the underlying networking and connectivity complexities of smart objects. And on the other hand, it provides a standardized application layer, primarily based on well-known and mature web protocols such as HTTP and REST, to promote interoperability and ease of use. In traditional IoT ecosystems, devices often rely on application, specific communication protocols, leading to tightly coupled systems and interoperability challenges, commonly referred to as silos of users. This limits the scalability and general applicability of IoT solutions.

The desired advantages of adopting the WoT approach are the simplification of application development. By reusing well-established web technologies, developers benefit from loose coupling, openness, and compatibility with a broad range of existing tools and infrastructures. In the WoT model, smart objects (or intermediaries acting on their behalf) expose their functionality through Web APIs, typically using RESTful interfaces. This allows other systems to interact with them using standard Web mechanisms, such as HTTP requests, thereby facilitating integration across heterogeneous platforms. The feasibility of this approach has increased significantly with advances in embedded systems, which now possess sufficient resources to support web-based communication stacks [6].

3.1 Abstract Architecture

The WoT is an architectural framework defined by the W3C to enable interoperability among IoT devices and services by using web-based technologies and principles. Its core objective is to bridge the heterogeneity gap among IoT ecosystems, allowing devices from different manufacturers, protocols, and networks to work together in a consistent, scalable, and secure way. As shown in Fig. 2.

The WoT Abstract Architecture, as shown in Fig. 3, provides a high-level model of how these systems are structured, describing the conceptual components, their roles, and how they interact through standardized interfaces and metadata descriptions. It does not dictate implementation specifics but ensures consistency and interoperability through well-defined conceptual roles.

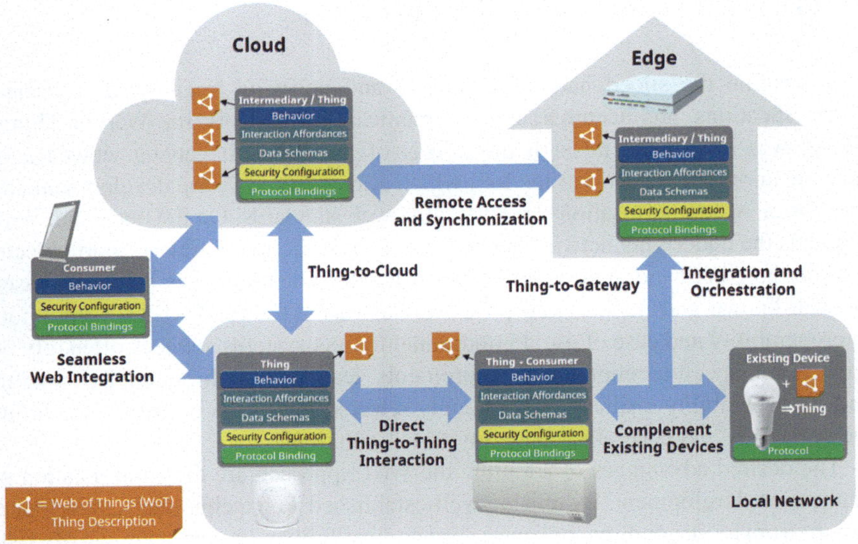

Fig. 2 Abstract architecture of W3C WoT [8]

3.2 Core Components of the WoT Abstract Architecture

3.2.1 Thing

At the center of the architecture is the Thing. It is a digital abstraction representing a physical or a virtual device that offers functionalities accessible via the web. A Thing can be a sensor, actuator, or service. In the WoT model, each Thing exposes its metadata and interactions via a standardized description known as a Thing Description (TD). The Thing can be either a physical object (e.g., a smart bulb) or a virtual resource (e.g., a database).

Interactions that a Thing exposes typically include:

- Properties: attributes that can be read and/or written (e.g., a temperature value, humidity, etc.)
- Actions: operations that can be invoked on the Thing (e.g., turning a device on or off)
- Events: asynchronous notifications about changes or conditions (e.g., a motion detection alert)

3.2.2 Thing Description (TD)

The Thing Description is a machine-readable metadata that provides a standardized description of a Thing's capabilities, interfaces, and operational constraints. It acts as a contract between the Thing and potential consumers, defining how the Thing

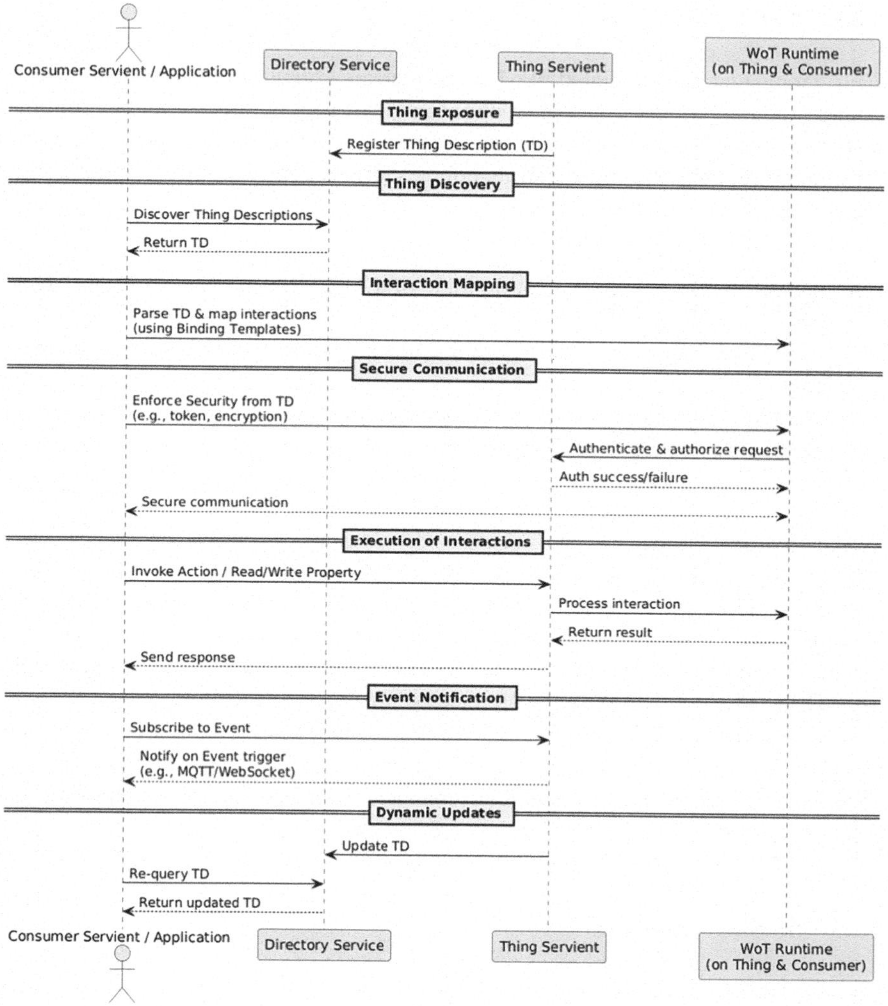

Fig. 3 WoT operational flow [8]

can be discovered, accessed, and interacted with. The TD includes metadata such as semantic metadata to describe what the Thing is; security configurations to specify access control and authentication mechanisms; and interaction affordances (Properties, Actions, Events) and how to access them over specific protocols (e.g., HTTP, CoAP, MQTT).

3.2.3 WoT Runtime

Each Thing has an associated WoT Runtime, an execution environment, which manages its operational logic, interactions, and security enforcement. The runtime enables a Thing to interpret and process requests, invoke actions, publish events, and enforce security and data handling policies based on its Thing Description. Such Things can also be sandboxed if needed.

The runtime is also responsible for:

- Exposing an API to the application logic of a Thing (or the behavior implementation)
- Protocol translation between external consumers and the Thing's native interfaces
- Interacting with external systems via network bindings
- Managing the life cycle of Things

3.2.4 Protocol Binding

Protocol Binding is essential because IoT ecosystems typically encompass a wide range of devices and networks, each with different resource constraints, communication patterns, and protocol stacks. By providing formal bindings to protocols like HTTP, MQTT, CoAP, and WebSockets, the WoT architecture abstracts away protocol-specific details while preserving the uniform interaction model described by TDs. This abstraction ensures that consumer applications can interact with Things in a consistent and predictable manner, regardless of the communication technology in use.

3.2.5 Binding Templates

Binding Templates define how the abstract affordances (Properties, Actions, Events) described in the Thing Description are concretely implemented using various network protocols and serialization formats. They map high-level interaction patterns to protocol-specific messages. For example:

- An Action might be mapped to a POST request over HTTP.
- A Property read might correspond to a GET request.
- An Event subscription might be handled via a WebSocket or MQTT topic.

Binding Templates decouple application logic from communication specifics, allowing the same application logic to interact with heterogeneous devices using different protocols transparently.

3.2.6 WoT Servient

A Servient is an entity that implements the WoT architecture and acts either as a Thing, a Consumer, or both. It hosts a WoT Runtime and provides a local execution environment for managing interactions.

- A Thing Servient exposes one or more Things with their Thing Descriptions.
- A Consumer Servient discovers Thing Descriptions, interprets them, and interacts with Things according to the affordances described.

Servients form the building blocks of WoT deployments, enabling devices, services, and applications to interact in a standardized manner.

3.3 Interactions and Workflow Within the Architecture

The operational flow within the WoT architecture is illustrated in the UML sequence diagram in Fig. 6 and typically proceeds through the following steps:

1. Thing Exposure: A Thing Servient generates a Thing Description that describes its capabilities and interaction affordances. This TD is made accessible through discovery mechanisms such as a directory service, broadcast, or direct access via a URL.
2. Thing Discovery: A Consumer Servient or application queries the available Thing Descriptions through discovery mechanisms. Once the TD is retrieved, it uses the metadata to understand how to interact with the Thing, including available operations and required security mechanisms.
3. Interaction Mapping: The Consumer Servient uses Binding Templates to map the abstract interactions (as defined in the TD) to concrete protocol requests. For example, to read a temperature property, the Consumer Servient might send an HTTP GET request to the URI specified in the TD.
4. Secure Communication: Before interactions can proceed, the WoT Runtime enforces the security configurations outlined in the TD. This may include token-based authentication, encryption, or other access control policies.
5. Execution of Interactions: Once authenticated and authorized, the Consumer Servient invokes Actions, reads or writes Properties, or subscribes to Events on the Thing. The WoT Runtime on the Thing Servient handles these requests, performing local processing as needed.
6. Event Notification: If a Consumer subscribes to an Event, the Thing Servient uses the appropriate protocol binding (e.g., MQTT publish, WebSocket push) to notify the Consumer when the Event occurs.
7. Dynamic Updates: In dynamic environments, Thing Descriptions can be updated, and Consumers can re-query or listen for TD updates to adapt to changes in a Thing's state or capabilities.

The WoT Abstract Architecture, as defined by W3C, provides a standardized, flexible, and secure framework for integrating IoT devices and services using web technologies. Its layered abstraction through Thing Descriptions, Binding Templates, and Servients allows for high interoperability while accommodating the protocol and operational heterogeneity inherent in IoT ecosystems.

By decoupling device capabilities from communication protocols and providing a uniform metadata-driven interaction model, the WoT architecture ensures that applications and services can seamlessly interact with a diverse range of Things—from simple sensors to complex cloud services—in a scalable, secure, and interoperable manner.

3.4 *Use Cases*

3.4.1 Use Case 1: Supply Chain Management

In a supply chain management context, interconnected logistics assets such as smart containers, transport vehicles, and warehouse devices expose their operational data through Thing Descriptions (TDs) within a Web of Things environment. As shown in Fig. 4, these descriptions may be registered in a WoT Directory to enable dynamic discovery by authorized logistics applications or accessed directly from the devices themselves. A centralized management platform, acting as a WoT Client Servient, retrieves TDs to interact with distributed resources. It reads operational parameters, monitors events such as shipment delays or environmental anomalies, and remotely triggers control actions. These interactions are facilitated via standardized protocol bindings (HTTP, CoAP, and MQTT) and protected through secure communication and access control mechanisms. This approach improves the visibility, responsiveness, and security of complex supply chain operations within a unified, web-based framework.

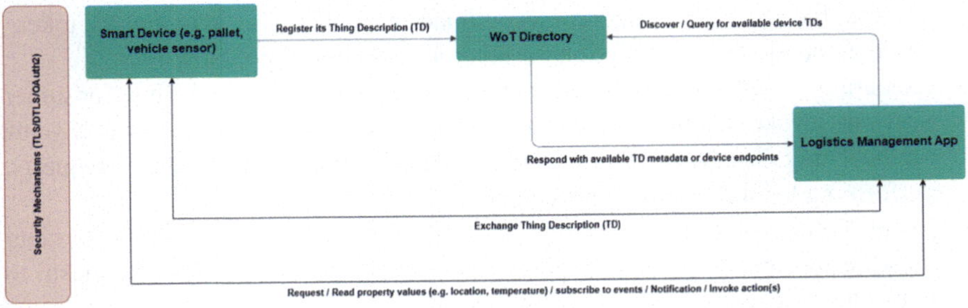

Fig. 4 WoT supply chain management uses case

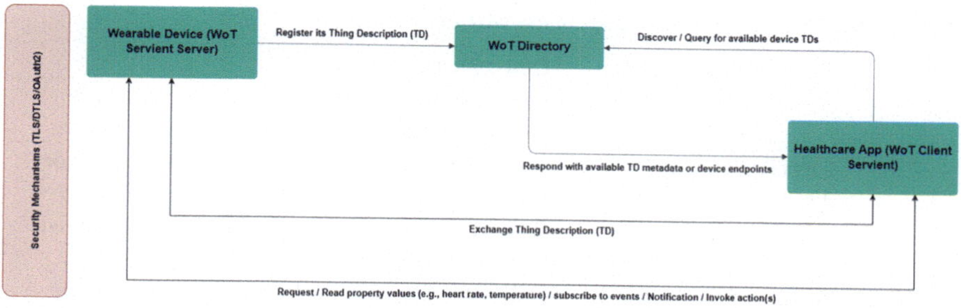

Fig. 5 WoT health monitoring uses case

3.4.2 Use Case 2: Healthcare Monitoring System

In a remote health monitoring scenario, wearable medical devices are integrated into the Web of Things ecosystem by describing their functionalities through Thing Descriptions (TDs), as shown in Fig. 5. These descriptions can be published to a WoT Directory or accessed directly by healthcare applications acting as WoT Client Servients. The application discovers available devices, retrieves their TDs, and performs secure interactions to collect patient data, manage device configurations, and receive alerts. Communication relies on WoT Protocol Bindings and is safeguarded by authentication, authorization, and encryption mechanisms to protect the confidentiality and integrity of sensitive health information. This model enables reliable and secure remote patient monitoring within a standardized, web-based framework.

As observed, although the two use cases share the same core architecture and interaction types, the nature, timing, scale, and criticality of these interactions differ according to domain-specific constraints. The WoT model is designed to abstract these communication patterns, but the implementation details and security measures must be adapted to the specific requirements of each use case.

4 Web of Things Security Threats

To better understand the security challenges related to the WoT, it is necessary to identify and categorize the potential threats.

These threats arise from the unique characteristics of the constrained IoT devices at physical level, network and communications level, and the application level. Table 2 below summarizes the most common and critical threats, encompassing aspects related to confidentiality, integrity, availability, application vulnerabilities, identity management, encryption, and physical security, extracted from [4, 5].

Table 2 Summary of common threats in the WoT environment

Threat ID	Threat name	Description
T1	Confidentiality Breaches	Unauthorized access to WoT devices, services, or data due to weak authentication or access controls
T2	Data Breaches and Leakage	Compromise leading to disclosure, modification, or theft of sensitive data generated by IoT devices
T3	Service Disruption (DoS)	Resource exhaustion through malicious or excessive requests, making services unavailable
T4	Misuse of Resources	Exploitation of device resources for unauthorized purposes like cryptomining or botnet activity
T5	Firmware/Software Tampering	Unauthorized modification of firmware or software to introduce malware or backdoors
T6	Privilege Escalation	Exploiting access control weaknesses to gain unauthorized elevated privileges
T7	Malicious Function Reconfiguration	Illicit modification of operational parameters or device functions
T8	Application Layer Threats	Exploiting application vulnerabilities such as insecure APIs, poor input validation, or weak session management
T9	Web-Based Attacks	Leveraging conventional web vulnerabilities (SQLi, XSS, CSRF) against WoT services and interfaces
T10	Identity & Access Management Weaknesses	Ineffective credential management, poor RBAC policies, or unrevised credentials allowing unauthorized access
T11	Insufficient E2E Encryption	Lack of secure, encrypted communication channels between devices and services, exposing data to interception
T12	Physical Attacks on Devices	Direct physical access enabling tampering, data extraction, or bypassing security mechanisms
T13	Side-Channel Attacks	Exploiting information leaked through power, timing, electromagnetic, or acoustic emissions
T14	Supply Chain Attacks	Insertion of malicious components during manufacturing, update, or distribution processes
T15	API Abuse	Exploitation of poorly secured, undocumented, or over-privileged APIs to bypass access controls or perform unauthorized actions
T16	Insecure Cloud Integration	Exploitation of cloud service vulnerabilities such as misconfigured permissions or weak API protections
T17	Weak or Missing Update Mechanisms	Lack of secure, authenticated, and verified firmware/software update processes
T18	Rogue or Cloned Devices	Introduction of unauthorized or malicious devices into the WoT environment
T19	Metadata Leakage	Exposure of metadata like device location or usage patterns enabling profiling or surveillance
T20	Network Reconnaissance	Scanning and mapping of WoT systems to identify active devices, services, or open ports
T21	Protocol Exploitation DoS	Attacks exploiting application layer protocols (e.g., CoAP, MQTT, WebSockets) to disrupt services

5 Architectural Approaches to WoT Security

Here we are looking at two different architectural ways to maintain security in small- or medium-scale WoT like home automation and large-scale WoT like Enterprise or Industry. The first approach relies on a smart gateway-based architecture, where gateways translate protocols between heterogeneous networks, such as sensor and IP networks, and serve as adaptable intermediaries for WoT security. The second approach is SDN-Based WoT (WoT/SDN) architecture, where SDN controllers manage network behavior to enforce security policies dynamically across devices. These are not mutually exclusive but rather explain the different directions of WoT security.

5.1 Smart Gateway-Based Architecture

As explained in [13], a WoT network is composed of two subnetworks: a sensor network and an IP network. In this context, a smart gateway in WoT is an adapter to receive, restructure, encapsulate, and retransmit protocol packets between the two subnetworks. Embedded Gateway (EG) is a gateway present in the WoT device itself and used for lightweight web protocols (UDP, CoAP). Independent Gateway (IG) is a separate physical device to convert protocols between heterogeneous networks to traditional web. A Communication Gateway (CG) generally converts protocols between a sensor network and an IP network, e.g., between ZigBee and IPv4. And Combined/Hybrid gateways perform multiple roles dynamically. The conceptual example is given in Fig. 6. The limitation of this architecture is that these gateways represent a single point of failure and that it is still a conceptual representation with no real implementation, evaluation, and no detailed information.

5.2 SDN-Based WoT (WoT/SDN) Architecture

According to [12] and as illustrated in Fig. 7, this architecture leverages the programmability of Software-Defined Networking (SDN) by integrating the resource layer of traditional resource-oriented web architectures with the SDN control plane. It incorporates dual SDN controllers, a Standard SDN Controller and a Security and Management Controller, and builds upon a previously proposed resource-oriented WoT system. Overall, the WoT/SDN architecture is organized into three layers.

The infrastructure layer is part of the WoT network, built using a multi-domain SDN (Software-Defined Networking) system. Sensors and actuators connect to gateways, which link to OpenFlow switches. Security tools like Deep Packet Inspection and Anti-DDoS systems can also connect directly to these switches. The SDN controller manages data flow—deciding whether to pass, block, or duplicate it—while monitoring network and security events through the OpenFlow protocol.

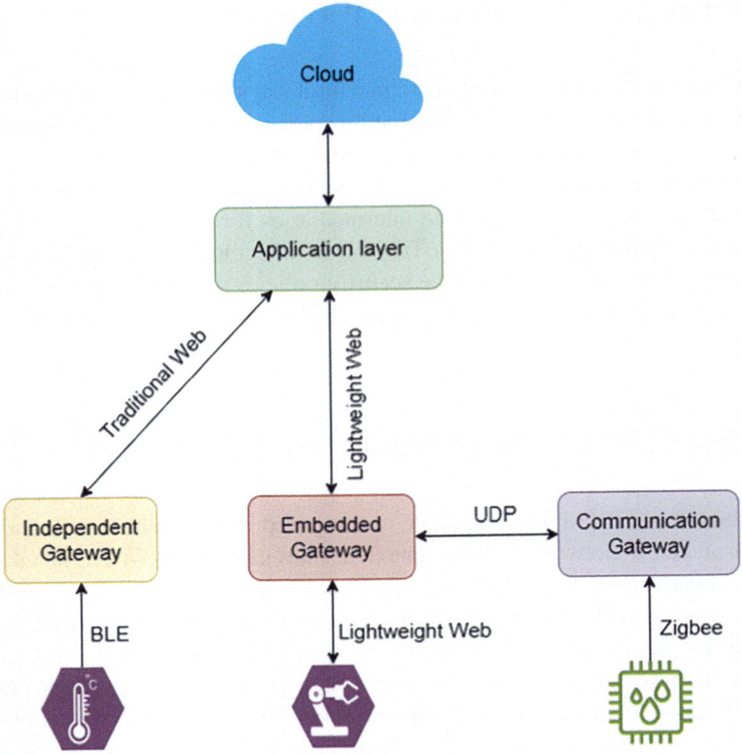

Fig. 6 Overview of smart gateway-based architecture

Resource coordination and control layer: This layer helps organize, store, and manage resources efficiently. It has two main functions: resource coordination and resource control. Resource coordination involves storing different types of data in a non-relational database, syncing them, listing shared resource details, and organizing them into resource pools for easy access. Resource control ensures smooth management of physical resources and security data. It uses SDN controllers—the Network Controller (NC) for network flow management and the Security and Management Controller (SMC) for handling security events and policies. With SDN integration, this layer improves resource coordination, security, and decision-making based on global network insights.

Application layer organizes and simplifies various resources into atomic services, which are reusable and universal. These services include resource subscription, device and application management, and policy services. Developers can use RESTful APIs to combine and customize these services to create new ones. Security services are broken down into basic functions like application identification, traffic analysis, and log analysis, making them more efficient and flexible. With SDN (Software-Defined Networking), security and application services can be restructured and optimized, improving agility, efficiency, and reducing overall costs.

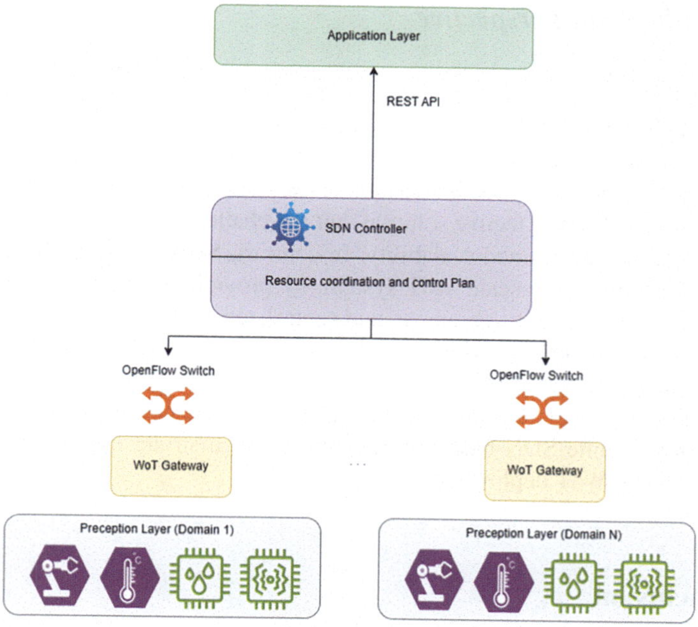

Fig. 7 Overview of SDN-based architecture

Table 3 Comparison between gateway and SDN-based WoT architectures

Category	Smart gateway-based architecture	SDN-based architecture
Security enforcement point	Edge level (near the device)	Network level (via centralized controller)
Primary function	Protocol transmission and lightweight access control	Centralized flow control, and dynamic policy enforcement.
Target devices	Legacy and resource-constrained devices	Large-scale IP-based networks with programmable nodes
Policy management	Static, locally configured	Dynamic, centrally managed via SMC
Real-time monitoring	Limited	Comprehensive, event-driven flow monitoring
Security services	Basic filtering, encryption, delegation, authentication	DoS mitigation, DPI, real-time anomaly detection, access control
Single point of failure risk	Gateway device	SDN controller

Another work, [7], explored an SDN-based architectural layer to dynamically enforce security policies across WoT environments, called smart spaces. By integrating SDN with WoT, it offers centralized control for access control management, network isolation, and dynamic management of security policies, particularly in smart city deployments. A comparison between these two paradigms is shown in Table 3.

5.3 Deployment Perspective

When creating a secure WoT architecture, the size and complexity of the deployment should be taken into consideration. Smart gateway-based architectures are best suited for small-scale or legacy environments that need basic security like encryption, access control, and protocol translation. The smart gateway-based architectures offer a simple and cost effective solution, but the challenge is having a single point of failure and having limited scalability. In contrast, SDN-based architectures are best for a dynamic large-scale WoT system, by providing programmable security policies, real-time traffic management, and centralized control. Because of this, they work better in commercial and industrial settings where flexibility, scalability, and exact control are essential.

In conclusion, smart gateway-based architecture works best for simple, localized configurations, while SDN-based architecture offers a strong foundation required for safe, scalable WoT deployments.

6 Conclusion

In conclusion, the WoT represents an important evolution in the IoT ecosystem by integrating heterogeneous IoT devices with web technologies, thereby enabling enhanced security, interoperability, and scalability. This chapter has provided a thorough overview of the WoT's abstract architecture, highlighting its core components and illustrating its practical relevance through two use cases. Central to this discussion has been the identification and analysis of the main security threats related to WoT environments. These threats underscore the complex and multilayered attack surface inherent to WoT systems. Moreover, understanding these risks is indispensable for developing robust and secure WoT architectures. The review of existing secure architectural solutions reveals the current strategies and gaps in safeguarding WoT deployments. However, there is still a need to further enhance the security of both IoT and WoT environments, as their increasing complexity and integration with critical infrastructures continue to expand the potential attack surface. In this context, blockchain technology emerges as a promising paradigm to address several WoT security challenges by providing decentralized trust, immutability, and enhanced transparency. The integration of blockchain with WoT security frameworks has the potential to significantly improve device authentication, data integrity, access control, and secure update mechanisms.

As the WoT continues to expand in scope and complexity, future research and development efforts must prioritize the design of holistic, scalable, and adaptable security models capable of addressing its diverse and evolving risks. Building on the foundation established in this chapter, the following sections of this book will first introduce the fundamental principles of blockchain technology, providing readers with the necessary background. The discussion will then progress to explore blockchain-based security solutions specifically designed to meet the unique requirements and constraints of WoT environments.

References

1. Lightweight Machine to Machine Technical Specification: Core. Open Mobile Alliance (2020), https://md2html-tool.com/docs/OpenMobileAlliance/LwM2M/master/c47850f/TS_Core/OMA-TS-LightweightM2M_Core-V1_2-20201110-A_full.html
2. OCF SPECIFICATION 2.2.8. OCF (2025), https://openconnectivity.org/developer/specifications/
3. AIOTI: High Level Architecture (HLA), Release 2.0, AIOTI WG03 – IoT Standardisation (2015)
4. Albarrak, K.M.: Securing the future of web-enabled iot: A critical analysis of web of things security. Applied Sciences **14**(23) (2024). https://doi.org/10.3390/app142310867, https://www.mdpi.com/2076-3417/14/23/10867
5. El Jaouhari, S., Bouabdallah, A., Bonnin, J.M.: Security issues of the web of things. In: Managing the web of things, pp. 389–424. Elsevier (2017)
6. Guinard, D., Trifa, V.: Building the Web of Things. Manning Publications Co (2016)
7. Jaouhari, S.E., Bouabdallah, A., Corici, A.A.: Sdn-based security management of multiple WoT smart spaces. J. Ambient Intell. Humaniz. Comput. **12**(10), 9081–9096 (2021). https://doi.org/10.1007/s12652-020-02601-y
8. Kajimoto, K., all: Web of things (WoT) architecture (Nov 2016), https://w3c.github.io/wot/architecture/wot-architecture.html
9. Nettsträter, A., al.: Internet of Things-Architecture IoT-A Deliverable D1. 3–Updated reference model for IoT v1. 5 (2012)
10. Sinha, S.: State of IoT 2024: Number of connected IoT devices growing 13% to 18.8 billion globally. IoT Analytics. (2024), https://iot-analytics.com/number-connected-iot-devices
11. Swetina, J., Lu, G., Jacobs, P., Ennesser, F., Song, J.: Toward a standardized common M2M service layer platform: Introduction to oneM2M. IEEE Wireless Communications **21**(3), 20–26 (June 2014)
12. Xiaofeng, Q., Wenmao, L., Teng, G., Xinxin, H., Xutao, W., Pengcheng, C.: WoT/SDN: web of things architecture using SDN. China Communications **12**(11), 1–11 (2015). https://doi.org/10.1109/CC.2015.7366240
13. Xie, W., Tang, Y., Chen, S., Zhang, Y., Gao, Y.: Security of web of things: A survey (short paper). In: Ogawa, K., Yoshioka, K. (eds.) Advances in Information and Computer Security. pp. 61–70. Springer International Publishing, Cham (2016)

Introduction to Blockchain Technology

Nour El Madhoun

1 Introduction

The emergence of Bitcoin has marked a revolution in digital transaction systems, introducing a decentralized model that eliminates the need for a Trusted Third Party (TTP). Thanks to a distributed ledger and peer-to-peer network, Bitcoin has addressed the problem of double spending by enabling transparent and verifiable exchanges in a trustless environment [1, 2]. The concept of blockchain predates Bitcoin, but it was the practical implementation of Bitcoin that triggered a new wave of innovation focused on decentralization and distributed trust.

After Bitcoin, blockchain technology has evolved into two main categories: Blockchain 1.0, which focuses on cryptocurrency transactions, and Blockchain 2.0, which introduced smart contracts that are programmable components enabling the automation of agreements on the network. The Ethereum blockchain is the prime example of this second generation [3, 4].

Since then, blockchain has extended far beyond financial systems. By integrating trust into cryptographic and algorithmic structures, it provides a basis for decentralized, secure, and transparent interactions. This potential has attracted interest from a variety of sectors, including voting, healthcare, identity management, supply chains [5], and also the Internet of Things (IoT) and the Web of Things (WoT), where blockchain enables data integrity, device autonomy, and traceability of interactions in distributed environments. However, effectively integrating blockchain into real-world applications requires careful consideration of decentralization capabilities, privacy trade-offs, and system constraints.

N. El Madhoun (✉)
LISITE Laboratory, Isep, Issy-les-Moulineaux, France
e-mail: nour.el-madhoun@isep.fr

S. El Jaouhari, N. El Madhoun (eds.), *Blockchain-based Web of Things Security: Advances, Challenges, and Solutions*, Signals and Communication Technology,
https://doi.org/10.1007/978-3-032-07671-7_2

This chapter introduces the main concepts of blockchain and explains how this technology can be used to implement a secure and decentralized infrastructure. It introduces the main consensus protocols, the structure and logic of smart contracts, and the main security properties of blockchain systems. Finally, it highlights some real-world applications of blockchain technology.

The chapter is structured as follows. Section 2 covers blockchain fundamentals. Section 3 presents consensus protocols. Section 4 introduces smart contracts. Section 5 examines security aspects. Section 6 highlights key applications. Section 7 concludes this chapter.

2 Blockchain Fundamentals

Blockchain is a distributed ledger technology that records transactions securely, transparently, and unfalsifiable. It works by organizing data into blocks, each containing a set of validated transactions and a cryptographic hash of the previous block, creating a continuous and immutable chain [1]. This structure ensures that once the data has been recorded, they cannot be modified retroactively without altering all subsequent blocks, which would require the consensus of the majority of the network.

Unlike traditional centralized databases, blockchain follows a peer-to-peer model in which all participating nodes maintain a synchronized copy of the ledger. When a transaction is initiated, it is broadcast over the network and verified collectively by the nodes. Once validated, the transaction is grouped with other transactions in a block and added to the chain according to a consensus protocol. This process guarantees both data consistency and resistance to manipulation, without relying on a TTP [2].

The architecture of blockchain systems can be seen in terms of functional layers. At the base is the data layer, which manages hashing, digital signatures, and Merkle tree structures. The network layer ensures communication and the propagation of transactions between peers. On top, the consensus layer reinforces agreement on the state of the ledger, even in the presence of untrusted or potentially malicious nodes [6]. Higher layers support advanced functions such as smart contracts and decentralized applications.

A key characteristic of blockchain technology is its ability to create trust between parties who do not know or trust each other. Trust is not based on institutional authority, but on cryptographic security, transparency, and the economic or computational cost of attacking the network [7]. This shift from trust in institutions to trust in algorithms has profound implications for the design of secure digital systems. In order to achieve this, the blockchain relies on consensus algorithms that determine how new blocks are validated and added to the chain. These mechanisms are essential to prevent inconsistencies, forks, and double-spending attacks. Their design has a significant impact on scalability, security, and energy consumption.

Blockchain systems are classified into several categories based on their access model and governance as follows [8, 9]: (1) public blockchains, such as Bitcoin and Ethereum, which are open to all and based on decentralized consensus without central control, (2) private blockchains, which limit participation to a known set of parties and are often used within organizations for internal record-keeping, (3) consortium blockchains are governed by a group of trusted organizations that share control; they are generally used in interorganizational contexts such as banking consortiums or supply chains, and (4) hybrid blockchains, which mix public and private elements to balance transparency, confidentiality, and efficiency. Each type meets different requirements in terms of scalability, trust, and compliance.

3 Consensus Mechanisms (Protocols)

The oldest and best-known consensus protocol is Proof of Work (PoW), introduced with Bitcoin [1]. In this protocol, the network participants (miners) compete to resolve a computationally intensive problem. The first to find a valid solution earns the right to add the next block to the chain and receives a reward. This mechanism secures the network by making it economically impossible for an attacker to take control of the majority of computing power (51%). However, PoW is known for its high energy consumption, limited throughput, and tendency to centralize mining in the hands of resource-rich actors [6].

As an alternative to the inefficiencies of PoW, the Proof of Stake (PoS) has been proposed. Instead of relying on computing power, PoS selects validators based on the amount of cryptocurrency they are willing to "stake" as a guarantee. The probability of being chosen to validate a block is proportional to the stake, which reduces energy consumption and improves efficiency. However, PoS introduces new challenges, such as the "nothing at stake" problem and the potential concentration of power among wealthy participants [4, 10]. Indeed, various alternative consensus protocols have emerged based on the PoS model:

- Proof of Burning (PoB): It requires validators to destroy irreversibly part of their cryptocurrency by sending it to an unusable address. This " burned " currency not only serves as proof of investment and commitment to the network but also introduces a financial risk that can discourage participation. The PoB has been explored in experimental systems such as SlimCoin [11].
- Proof of Capacity (PoC): It is also known as Proof of Space. It allows validators to provide unused disk space rather than computing power. The participants pre-calculate and store possible solutions to the consensus puzzle, and the more storage space they allocate, the more likely they are to produce a valid block. PoC is more energy-efficient and can run on standard consumer hardware, but it is susceptible to storage-based attacks and implementation complexity [12, 13].
- Proof of Elapsed Time (PoET): It was proposed by Intel and implemented in Hyperledger Sawtooth. It assigns block creation rights based on a randomized

waiting period, with the shortest wait time winning the right to produce the next block. The randomness and fairness of this method depend on Trusted Execution Environments (TEEs), which raises concerns about hardware reliance and trust assumptions [6, 14].

Each of these protocols offers different trade-offs in terms of security, decentralization, energy efficiency, and economic incentives. Comparisons based on tables often not only highlight the reduced energy consumption of PoS and PoC compared with PoW but also note the risks of centralization introduced by issue-based selection or the vulnerability of storage-intensive protocols.

Other consensus mechanisms have also become increasingly popular in different blockchain architectures:

- Delegated Proof of Stake (DPoS): It improves scalability by allowing token holders to elect a small group of delegates to validate transactions and produce blocks. Although it offers higher throughput, it can reduce decentralization due to the limited number of validators [15].
- Practical Byzantine Fault Tolerance (PBFT) and its derivatives: They are commonly used in permissioned blockchains. These protocols enable nodes to reach consensus even if some participants act maliciously, thanks to multiple rounds of voting. PBFT guarantees finality and low latency but is less scalable for large decentralized networks [16].
- Proof of Authority (PoA): It is used in private or consortium blockchains and is based on a fixed set of trusted validators whose identities are known and verified. PoA offers high efficiency but is centralized by design and therefore less suited to untrusted public environments [17].

The choice of consensus protocol determines the performance, security guarantees, and governance model of the blockchain. The next section introduces smart contracts and explains how they enable programmable and autonomous behavior on top of the distributed ledger.

4 Smart Contracts

Smart contracts are one of the major evolutions introduced in blockchain systems beyond simple value transfers. The concept was first formalized by Nick Szabo in 1994 as a set of digital promises, automatically executed under predefined conditions and without the need for human intervention or trusted intermediaries. In the context of blockchain, smart contracts are deployed on the blockchain and function in a decentralized, transparent, and tamper-proof environment [18].

A smart contract is primarily a program integrated into the blockchain that is initiated by transactions. When specific conditions are met, which are usually defined by the contract logic, it executes autonomously. The program can store data, interact with other contracts, and generate results such as asset transfers or status changes. Once deployed, its code becomes immutable: Any modification requires the deployment of a new contract referencing the previous one, which guarantees traceability and auditability over time [19].

Compared with traditional software on centralized servers, smart contracts benefit from all the properties of the blockchain, including decentralization, determinism, and verifiability. Each node in the network can read the contract code and independently validate its execution. This transparency eliminates the need for a TTP and allows participants to rely on the algorithm itself as a source of trust [20]. A standard smart contract includes variables (to store data), functions (to define logic), and a state (to reflect the current context of execution). When a smart contract is initiated, either by a user or by another contract, it processes the input, updates its internal state, and records the result in the blockchain. These steps are visible publicly, not only enhancing trust but also raising concerns about privacy and confidentiality. Because blockchain ledgers are transparent by design, all contractual data, including sensitive information, can be accessed by all participants [6].

The integration of smart contracts makes it possible to automate complex workflows, such as financial agreements, governance mechanisms, and asset management. Their use has given rise to decentralized Applications (dApps) and entire decentralized ecosystems such as Decentralized Finance (DeFi). However, the writing of secure smart contracts remains a major challenge, as bugs or vulnerabilities in the code can have a potentially irreversible impact. The immutability of deployed contracts means that errors cannot be easily corrected, underlining the need for formal verification and strict auditing practices [4]. With decentralized logic, trustless automation, and new forms of digital interaction, smart contracts transform blockchain from a passive storage system into an active, programmable infrastructure.

5 Security Aspects

One of the fundamental promises of blockchain technology is its security. Its architecture is designed to guarantee data integrity, availability, and traceability without relying on centralized control. In general terms, blockchain security can be analyzed at two levels: security at the infrastructure level and security at the transaction level [6, 21–24]:

1. Security at the infrastructure level: At the infrastructure level, several key properties contribute to the resilience of blockchain.
   - Decentralization eliminates single points of failure by distributing control and data storage between all participating nodes. The result is a system that is resistant to targeted attacks against any central authority or database.
   - Transparency ensures that all participants can independently verify transactions. While user identities are masked by cryptographic addresses, the

transaction data itself is visible to all, which contributes to verifiability but may raise confidentiality issues in some use cases.

- Immutability is guaranteed thanks to the cryptographic chaining of blocks. Attempting to modify a block would require the recalculation of all subsequent blocks and would take control of the majority of the network's consensus power. This means that once data has been validated and added to the chain, they cannot be retroactively modified.
- Consistency means that all honest nodes share a consistent vision of the ledger. The level of consistency depends on the consensus algorithm used. For example, the PoW algorithm provides eventual consistency, where finality is probabilistic and improves over time, while some PoS variants may offer faster or deterministic finality.
- Despite these advantages, blockchains remain vulnerable to specific security threats:
    - Distributed Denial of Service (DDoS) attacks can be targeted at nodes or the network itself, which can disrupt transaction propagation or block validation.
    - Double-Spending attacks try to spend the same digital asset more than once by exploiting temporary inconsistencies in the ledger, particularly in systems with probabilistic finality.
    - 51% attacks happen when a single entity or coalition gains majority control of the network's consensus power. This enables the attacker to modify transaction histories, prevent confirmations, or censor certain participants. Even though such attacks are possible in theory, their cost and complexity increase with the size and distribution of the network.

2. Security at the transaction level: At the transaction level, blockchain offers several layers of protection:
    - Authentication is achieved thanks to digital signatures, which confirm the origin and integrity of each transaction.
    - Anonymity, or more precisely pseudonymity, masks the identities of users entirely within public keys. However, link analysis or metadata correlation can challenge this principle, particularly in transparent ledgers such as Bitcoin.
    - Confidentiality is not a natural characteristic of most public blockchains. While the transaction is secure and inviolable, its content remains visible to all participants. This limits the use of blockchain in scenarios involving sensitive personal or business data, unless reinforced by additional cryptographic techniques such as Zero-Knowledge Proofs.

6 Applications

Blockchain technology, initially used for the first time in cryptocurrencies, has been applied in a very wide range of areas where trust, transparency, and verifiability are crucial. In e-voting, it guarantees the integrity and auditability of ballots while preserving the anonymity of voters [25]. In healthcare, it enables patients to control access to their medical data via smart contracts, despite the fact that full blockchain storage remains impractical due to issues of confidentiality and scalability [26]. For identity and access management, decentralized identity frameworks allow individuals to selectively own and share their credentials without depending on central authorities [27, 28]. Blockchain also supports access control in IoT environments by recording and managing permissions via smart contracts, improving traceability while also raising privacy concerns [29].

In supply chain management, blockchain technology creates immutable records for the traceability of goods from origin to delivery by reducing fraud and improving accountability [30]. The notarization services supported by blockchain time-stamp and prove the existence and integrity of digital assets, which is especially useful for the protection of intellectual property [31]. In addition, the successful integration of blockchain with IoT and WoT facilitates secure communication between machines and decentralized coordination. Blockchain adds trusted and auditable data to IoT, whereas WoT relies on web standards to make connected devices interoperable on infrastructures based on blockchain.

These various applications demonstrate the evolution of blockchain from a financial innovation to a fundamental technology for building trusted digital ecosystems. However, their deployment is still required to address limitations related to confidentiality, scalability, interoperability, and regulatory compliance.

7 Conclusion

Blockchain is redefining digital trust thanks to cryptography and decentralization. This chapter has presented its basic architecture, consensus protocols, smart contracts, security principles, and key application areas. Beyond cryptocurrencies, blockchain offers a reliable basis for secure, transparent, and decentralized systems.

References

1. Satoshi Nakamoto. Bitcoin: A peer-to-peer electronic cash system, 2008.
2. Arvind Narayanan and Jeremy Clark. Bitcoin's academic pedigree. *Communications of the ACM*, 60(12):36–45, Nov 2017.
3. Damiano Di Francesco Maesa and Paolo Mori. Blockchain 3.0 applications survey. *Journal of Parallel and Distributed Computing*, 138:99–114, Apr 2020.

4. Gavin Wood. Ehtereum: A Secure Decentralized Generalized Transaction Ledger. *Ethereum Project Yellow Paper*, pages 1–32, 2018.
5. Nour El Madhoun, Julien Hatin, and Emmanuel Bertin. A decision tree for building it applications. *Annals of Telecommunications*, 76(3):131–144, 2021.
6. Rui Zhang, Rui Xue, and Ling Liu. Security and Privacy on Blockchain. *ACM Computing Surveys*, 52(3):1–34, Jul 2019.
7. Lakshmi Siva Sankar, M Sindhu, and M Sethumadhavan. Survey of consensus protocols on blockchain applications. *2017 4th International Conference on Advanced Computing and Communication Systems (ICACCS)*, pages 1–5, 2017.
8. Sana Zeba, Preetam Suman, and Kanishka Tyagi. Types of blockchain. In *Distributed Computing to blockchain*, pages 55–68. Elsevier, 2023.
9. Harsh Sheth and Janvi Dattani. Overview of blockchain technology. *Asian Journal For Convergence In Technology (AJCT) ISSN-2350-1146*, 2019.
10. Moneeb Ahmed, Ihsan Elahi, Muhammad Abrar, Umair Aslam, Iqra Khalid, and Muhammad Asif Habib. Understanding Blockchain. *Proceedings of the 3rd International Conference on Future Networks and Distributed Systems - ICFNDS '19*, pages 1–8, 2019.
11. Kostis Karantias, Aggelos Kiayias, and Dionysis Zindros. Proof-of-Burn. In Joseph Bonneau, and Nadia Heninger, editors, *Financial Cryptography and Data Security*, pages 523–540. Springer International Publishing, 2020.
12. Shehar Bano, Alberto Sonnino, Mustafa Al-Bassam, Sarah Azouvi, Patrick McCorry, Sarah Meiklejohn, and George Danezis. SoK: Consensus in the age of blockchains. In *Proceedings of the 1st ACM Conference on Advances in Financial Technologies*, AFT '19, page 183–198, New York, NY, USA, 2019. Association for Computing Machinery.
13. Stefan Dziembowski, Sebastian Faust, Vladimir Kolmogorov, and Krzysztof Pietrzak. Proofs of space. In Rosario Gennaro and Matthew Robshaw, editors, *Advances in Cryptology – CRYPTO 2015*, pages 585–605, Berlin, Heidelberg, 2015. Springer Berlin Heidelberg.
14. Sawtooth. Hyperledger Sawtooth, 2016. [Retrieved: 28-03-2022].
15. Qian Hu, Biwei Yan, Yubing Han, and Jiguo Yu. An improved delegated proof of stake consensus algorithm. *Procedia Computer Science*, 187:341–346, 2021.
16. Nikos Chondros, Konstantinos Kokordelis, and Mema Roussopoulos. On the practicality of practical byzantine fault tolerance. In *ACM/IFIP/USENIX International Conference on Distributed Systems Platforms and Open Distributed Processing*, pages 436–455. Springer, 2012.
17. Manuel Adelin Manolache, Sergiu Manolache, and Nicolae Tapus. Decision making using the blockchain proof of authority consensus. *Procedia Computer Science*, 199:580–588, 2022.
18. Maria G Vigliotti. What do we mean by smart contracts? Open challenges in smart contracts. *Frontiers in Blockchain*, 3:553671, 2021.
19. Bhabendu Kumar Mohanta, Soumyashree S Panda, and Debasish Jena. An Overview of Smart Contract and Use Cases in Blockchain Technology. *2018 9th International Conference on Computing, Communication and Networking Technologies (ICCCNT)*, pages 1–4, Jul 2018.
20. Andreas M Antonopoulos and Gavin Wood. *Mastering Ethereum*. O'Reilly Media, first edition, 2018.
21. David Shrier, Weige Wu, and Alex Pentland. Blockchain & infrastructure (identity, data security). *Massachusetts Institute of Technology-Connection Science*, 1(3):1–19, 2016.
22. Thomas Hardjono and Ned Smith. Decentralized trusted computing base for blockchain infrastructure security. *Frontiers in Blockchain*, 2:24, 2019.
23. Daniel Maldonado-Ruiz, Jenny Torres, and Nour El Madhoun. Fundamentals of blockchain technology. In *Principles and practice of blockchains*, pages 3–25. Springer, 2022.
24. Xiaoqi Li, Peng Jiang, Ting Chen, Xiapu Luo, and Qiaoyan Wen. A survey on the security of blockchain systems. *Future Generation Computer Systems*, 107:841–853, Jun 2020.
25. Ali Benabdallah, Antoine Audras, Louis Coudert, Nour El Madhoun, and Mohamad Badra. Analysis of blockchain solutions for e-voting: a systematic literature review. *IEEE Access*, 10:70746–70759, 2022.

26. Darine Al-Mohtar, Amani Ramzi Daou, Nour El Madhoun, and Rachad Maallawi. A secure blockchain-based architecture for the covid-19 data network. In *2021 5th Cyber Security in Networking Conference (CSNet)*, pages 1–5. IEEE, 2021.
27. Daniel Maldonado-Ruiz, Alan Pulval-Dady, Yulin Shi, Zhe Wang, Nour El Madhoun, and Jenny Torres. NestedChain: "blockchain-inside-a-blockchain" new generation prototype. *Annals of Telecommunications*, 79(11):881–899, 2024.
28. Daniel Maldonado-Ruiz, Chan Yeong Hwang, Elina Jankovskaja, Karima Sadykova, Messilva Mazari, Jenny Torres, and Nour El Madhoun. Implementation of two-signature security on an identity storage smart contract. In *The 39th International Conference on Advanced Information Networking and Applications (AINA-2025)*, 2025.
29. Saad El Jaouhari, Ahmed Bouabdallah, and Andreea Ancuta Corici. SDN-based security management of multiple wot smart spaces. *Journal of Ambient Intelligence and Humanized Computing*, 12(10):9081–9096, 2021.
30. Nir Kshetri. *Blockchain and supply chain management*. Elsevier, 2025.
31. Athina-Styliani Kleinaki, Petros Mytis-Gkometh, George Drosatos, Pavlos S Efraimidis, and Eleni Kaldoudi. A blockchain-based notarization service for biomedical knowledge retrieval. *Computational and structural biotechnology journal*, 16:288–297, 2018.

Blockchain-Based Approaches for Securing The Web of Things (WoT)

Saad El Jaouhari

1 Introduction

The evolution of the Internet of Things (IoT) has significantly enhanced how devices interact and communicate, fostering smarter and more efficient environments. The global number of IoT devices is expected to reach 40.6 billion by 2034 [1]. These devices continuously generate and exchange data, offering opportunities for automation, optimization, and real-time decision-making. However, the rapid growth and heterogeneity of IoT systems have also introduced significant security and reliability challenges. IoT devices are often resource-constrained, deployed in untrusted environments, and lack standardized security mechanisms. As a result, they are vulnerable to a wide range of threats, including sensor faults, data tampering, (D)DoS attacks, and unauthorized access.

To address these limitations, the Web of Things (WoT) emerges as a significant advancement by building on IoT and integrating web technologies such as HTTP, WebSockets, and RESTful APIs. By standardizing communication protocols and enabling seamless device integration, WoT not only improves interoperability but also facilitates centralized management and monitoring, which are essential for enhancing security and reliability in complex IoT ecosystems. It also brings the maturity of well-established web-based security techniques into the WoT domain, offering mechanisms such as HTTPS encryption, OAuth-based authentication, and secure API design that can be leveraged to protect device interactions and data exchanges. These technologies contribute to a more structured and scalable approach to securing interconnected devices, especially in environments where traditional IoT security measures fall short. Nevertheless, WoT systems remain

S. El Jaouhari (✉)
Isep, LISITE, Issy-les-Moulineaux, France
e-mail: saad.el-jaouhari@isep.fr

S. El Jaouhari, N. El Madhoun (eds.), *Blockchain-based Web of Things Security: Advances, Challenges, and Solutions*, Signals and Communication Technology,
https://doi.org/10.1007/978-3-032-07671-7_3

vulnerable to a range of cyber threats, including unauthorized access, data breaches, and Denial-of-Service (DoS) attacks, highlighting the need for additional security layers and continuous monitoring [2].

On the other hand, blockchain technology, with its intrinsic characteristics of decentralization, immutability, cryptographic security, smart contracts, and transparency, offers a promising solution to enhance the security of WoT and IoT environments. It introduces innovative capabilities in areas such as device identity management, secure data sharing, and decentralized access control, effectively addressing several critical vulnerabilities inherent in WoT systems. By eliminating the need for a central authority, blockchain reduces single points of failure and enhances overall system resilience. Its cryptographic foundations ensure data integrity and confidentiality, while smart contracts enable automated, secure transactions and fine-grained access control [3]. These features are essential for protecting WoT ecosystems against cyber threats such as unauthorized access, data tampering, and DoS attacks [4]. As a result, blockchain provides a robust and trustworthy foundation for securing interactions and data exchanges in increasingly complex and distributed WoT networks. Therefore, the main objective of this chapter is to highlight the key contributions of using blockchain technology to secure the WoT through a comprehensive analysis of the existing literature.

This chapter is organized as follows: Sect. 2 introduces the security requirements of the WoT. Section 3 provides an overview of blockchain-based security solutions for securing WoT environments. Section 4 discusses the main remaining security challenges, and the chapter concludes in Sect. 5.

2 WoT Main Security Issues

As highlighted in the previous chapter, the WoT introduces a unifying layer that leverages standard web protocols such as HTTP, REST APIs, and JSON. This allows devices to communicate seamlessly, much like how applications interact over the Internet. On the web, communication is governed by well-defined application layer protocols such as HTTP, HTTPS, FTP, and SMTP, which specify rules and formats for data exchange. For example, devices understand that an HTTP request includes a method like "GET" or "POST," headers, and an optional body. By adopting these standards, WoT enables direct communication between devices, such as a Nest Thermostat and a Philips Hue bulb, without the need for intermediary middleware. However, this integration introduces consequential security and privacy protection challenges. The open nature of WoT systems expands the attack surface, exposing devices to various vulnerabilities.

2.1 Threats

According to the World Wide Web Consortium (W3C) [5], the WoT faces numerous security threats due to its open interfaces, reliance on network communications, and potential for code execution inside runtime environments. These threats affect the integrity, confidentiality, authenticity, and availability of system components, user data, and Thing Descriptions (TDs). Attackers can range from remote network attackers to insiders and malicious system providers. W3C's threat model emphasizes securing interfaces, runtime environments, and communication channels, especially when multiple tenants or dynamic scripting are involved, as resumed in Table 1.

Moreover, WoT devices commonly use web-based authentication protocols like OAuth 2.0 and OpenID Connect, which, despite their widespread adoption, are vulnerable to token theft and session hijacking. Additionally, exposed APIs in WoT systems can be exploited through weak authentication, authorization flaws, and injection attacks. Additional challenges include secure access control, device authentication/identity management, device management, data protection/privacy respect, trust management resilience against cyberattacks, and scalability of security mechanisms to accommodate the dynamic and large-scale nature of WoT deployments [6].

While WoT inherits many traditional IoT security threats, it also introduces additional risks due to its reliance on web technologies. The authors in [7] provide a comparative analysis of the main security threats in IoT and WoT environments. In authentication and authorization, IoT suffers from weak credentials and lack of multifactor authentication, while WoT faces vulnerabilities in OAuth and OpenID Connect, including token theft and session hijacking.

Data security in WoT is challenged by real-time data sharing and exposed APIs, increasing the risk of interception and manipulation. Network security threats expand in WoT due to exposure to web-based attacks like SQL injection and XSS and more complex communication patterns. Software security concerns shift from outdated firmware in IoT to web application vulnerabilities and insecure third-party components in WoT. Physical security remains a concern, but WoT devices are more prone to being used as attack vectors due to their web connectivity. Finally, data privacy risks are amplified in WoT by real-time data flows and regulatory compliance challenges, such as GDPR and CCPA.

Also the authors in [2] provided a critical analysis of the WoT security landscape, focusing on the vulnerabilities introduced by integrating web technologies into IoT systems. Using a threat modeling approach, the paper identifies key threats, their attack vectors, affected components, and exploited vulnerabilities, supported by misuse patterns to illustrate attacker behavior. It also proposes tailored defenses and introduces UML-based architectural models to visualize system interactions and risks, along with a simulated attack scenario to demonstrate real-world implications and the need for robust security strategies.

The authors in [8] proposed Zion, an open-source, scalable W3C Thing Description Directory (TDD) designed for efficient indexing and querying of IoT devices

Table 1 Summary of WoT security threats (based on W3C guidelines) [5]

Threat category	Threat description	Main targets	Attack methods
Protocol Binding Exploit	Exploiting protocol binding interfaces to compromise internal components	WoT assets within same execution boundary	Remote interface abuse or compromised protocol binding component
Interface Exposure	Unauthorized access to or misuse of exposed WoT interfaces	Devices, user data, TDs	Compromised Exposed Thing or Interface
Interface DoS	Disruption of services by overwhelming WoT interfaces	Infrastructure and Thing Resources	Message flooding, excessive resource consumption
TD Authenticity Attack	Modifying Thing Descriptions during transit	TDs in-transfer	MITM attack, rollback to older TD versions
TD Confidentiality Breach	Intercepting private fields or metadata of TDs	Confidential TD data	Network sniffing, passive observation
User Data Authenticity Attack	Alteration of user data in transit	System user data	Data replay, injection, tampering
User Data Confidentiality Breach	Unauthorized access to user data during transmission	Private user data	Network eavesdropping, unsecured transport
Side-Channel Analysis	Behavioral pattern inference from message observation	WoT behavior metrics	Passive traffic analysis
Script Installation Threat	Insertion of malicious scripts into WoT runtime	WoT runtime environment	Exploiting script management API
Runtime Compromise	Full control over the WoT runtime gained by attacker	Runtime and all hosted assets	Runtime privilege escalation, compromised API
Script-Level DoS	Disruption of WoT service via destructive scripts	Devices and infrastructure	Script deletion, resource exhaustion
Untrusted Script	Malicious script interferes with others in the same runtime	Scripts, configurations, user data	Inter-script attacks, isolation failure
TD Tampering	Modification of stored TDs via scripts	Stored TDs on devices	Local rollback or unauthorized modification
User Data Manipulation via Script	Script modifies or fabricates user data locally	Stored or processed user data	Tampering by cohosted or rogue scripts
Confidentiality Breach via Script	Accessing data of other scripts from within shared runtime	Internal configurations, script data	Unauthorized read or memory scraping
Multi-Tenant DoS	Resource abuse by one tenant impacting others	Coexisting system provider resources	Local resource overuse, memory/CPU starvation

using JSONPath. It supports full CRUDL operations for authenticated users, real-time event subscriptions, and token-based authentication. Performance evaluations show Zion outperforms other TDDs like WoT Hive and TinyIoT, maintaining linear scalability under high workloads and supporting future extensions like geospatial queries and tagging.

Addressing these multifaceted issues requires robust, lightweight, and interoperable security solutions tailored to the unique constraints and requirements of WoT systems.

2.2 *Privacy Consideration*

The WoT introduces several privacy risks that must be addressed during system design and deployment. Key risks include the disclosure of device and system configurations through TDs and network interactions, leakage of user data via intermediaries or public interfaces, and the potential for user tracking through persistent identifiers or system fingerprinting. To mitigate these risks, the W3C guidelines recommend minimizing publicly exposed information, securing communications with encryption, limiting access to TDs based on client privileges, and periodically rotating identifiers. Additionally, users should be informed about data collection practices and given control over their privacy exposure, aligning with principles of data minimization, confidentiality, and user participation [5].

3 State of the Art

This section highlights the key blockchain-based solutions proposed in the literature to enhance the security of the WoT.

For instance, the authors in [9] proposed a secure and efficient architecture for WoT systems by leveraging digital twins and permissioned blockchains. Instead of direct access to IoT devices, consumers interact with smart contract-based digital twins deployed on a Hyperledger Fabric network. These digital twins conform to W3C WoT standards, ensuring interoperability and abstraction. The architecture enhances security through decentralized identity management, immutability, and auditability, while also improving availability and reliability. The system includes a setup phase for device registration and a secure access phase where authenticated users interact with devices via the blockchain. A proof-of-concept implementation demonstrates the feasibility of the approach using a smart lamp and WoT-compliant gateways.

This system [10] used blockchain to link mobile databases—each embedded in devices like Raspberry Pi or UP Board—as blocks storing local and network-wide sensor data. Built on a WoT-based web server, it enables secure, decentralized data collection, visualization, and cloud access. Each node aggregates and replicates sensor data across the network, ensuring integrity and availability. AI-driven blockchain management and eMMC-based storage enhance performance, while eliminating mining overhead and supporting real-time, browser-based access.

This work [11] presented a four-layer SHM architecture integrating W3C WoT for interoperability and Ethereum blockchain for data integrity and traceability. Critical sensor data are filtered and published via a Smart Contract by a third-party SHM Controller. Built on the MODRON platform, the system ensures secure, decentralized access to structural health data. Performance tests show the Smart Contract scales efficiently under load, with reduced latency and gas costs using the ERC-1967 proxy pattern.

The work in [12] proposed a decentralized identity and trust management framework for IoT using Self-Sovereign Identities (SSIs) and a Web of Trust model built on the IOTA Tangle. Devices self-assign identities based on W3C DIDs, enabling permissionless authentication and trust scoring without a central authority. The system ensures tamper-proof identity claims, low computational overhead, and high scalability. A proof-of-concept demonstrates feasibility and resilience against common IdMS and WoT attack vectors.

The authors in [13] presented a blockchain-based framework for securing IoT/WoT data in large-scale, multi-organization environments. It formalizes data certification and integrity for heterogeneous models and time series, automates verification, and supports flexible efficient certification. Implemented on Hyperledger Fabric and integrated with the Snap4City platform, the system ensures traceability, sustainability, and performance in mobility applications. Validation includes performance assessment and deployment in the CN MOST OPTIFaaS initiative.

This paper [14] proposed a scalable, decentralized PKI for IoT using blockchain and web security standards. It replaces centralized certificate authorities with a Web of Trust model, storing trust relationships on Ethereum while keeping certificates off-chain. The system supports cross-organizational authentication via ACE framework and enables secure public key distribution without trusted intermediaries. A prototype demonstrates feasibility and outlines future work on privacy and provisioning.

SemChain is a lightweight, CoAP-based framework integrating Semantic Web, IoT, and permissioned blockchain to enhance secure, interoperable resource discovery presented in [15]. It uses distributed resource directories as blockchain peers and smart contracts for semantic registration and querying. Two storage models, SemChain-Full and SemChain-Hash, are evaluated in a smart hospital use case, showing improved precision, recall, and scalability while addressing the limitations of centralized and resource-heavy IoT systems.

This paper [16] presented a WoT-oriented distributed storage framework using IOTA and MQTT/MQTT-SN for secure, low-power M2M communication. It supports geographically distributed IoT sites with fog computing and IOTA channels for immutable event logging. The architecture includes granular access policies, modular components, and a custom protocol. Simulations and qualitative analysis show strong performance in latency, throughput, scalability, and mobility, making it suitable for intelligent, decentralized IoT ecosystems.

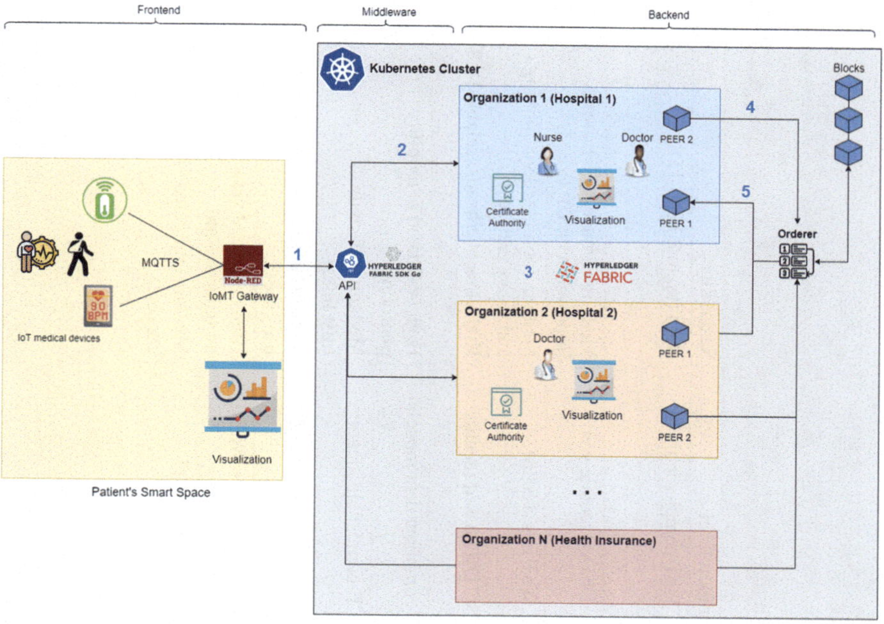

Fig. 1 Blockchain-based architecture for securing IoT health data from the patient's smart space [17]

The authors in [17] presented a secure, permissioned blockchain framework for the Internet of Medical Things (IoMT) using Hyperledger Fabric. It ensures privacy, confidentiality, integrity, and traceability of sensitive health data collected via IoMT sensors. The system uses Node-RED for device integration and X.509 certificates for identity management. A proof of concept demonstrates real-time data exchange, ledger immutability, and verifiable transaction history. Future work includes extending smart contract logic to broader healthcare stakeholders. The architecture of the proposed solution is shown in Fig. 1.

Furthermore, this architecture can be tailored to follow the W3C WoT guidelines. Each IoMT device, such as wearable or implantable sensors in the patient smart space, is modeled as a W3C WoT "Thing," described using a "Thing Description" (TD) that semantically defines its properties (e.g., heart rate), actions (e.g., startMeasurement), and events (e.g., fallDetected). These TDs allow seamless discovery and interoperability, enabling the local coordinator or gateway to act as a "WoT Servient" that dynamically identifies nearby Things and interacts with them regardless of underlying communication protocols. Through this abstraction, the gateway can invoke device functions, monitor real-time events, and translate data across heterogeneous devices. Additionally, the WoT Servient interfaces with the Hyperledger Fabric API to ensure secure data exchange and immutable storage on a blockchain infrastructure, guaranteeing data integrity, traceability, and authorized access control in the healthcare context. A comparison of these works is shown in Table 2.

Table 2 Comparison of state-of-the-art works on WoT security using blockchain

Ref.	Blockchain type	Security features	WoT integration level	Scalability
[9]	Permissioned (Hyperledger Fabric)	Authentication, Integrity, Auditability, Availability, Access Control	Full (W3C WoT-compliant digital twins)	High (up to 20,000 TPS)
[10]	Public (Bitcoin-like)	Integrity, Confidentiality, Decentralization, Redundancy	Partial (WoT-based mobile web server)	Medium (limited by hardware, improved with UP Board)
[11]	Public (Ethereum)	Integrity, Traceability, Auditability	Full (W3C WoT-based SHM pipeline)	Medium (optimized via filtering and proxy pattern)
[12]	Public/Permissionless (IOTA Tangle)	SSI, Trust Scoring, Integrity, Decentralization	Partial (W3C DID-based identity layer)	High (low overhead, scalable DLT)
[13]	Permissioned (Hyperledger Fabric)	Data Integrity, Certification, Automation, Multi-org. Trust	Full (WoT-compliant data models)	High (validated in mobility platforms)
[14]	Public (Ethereum)	Authentication, Decentralized Trust, PKI Replacement	Partial (ACE-compatible identity layer)	Medium (off-chain certs, on-chain trust links)
[15]	Permissioned (Smart Contracts)	Integrity, Trust, Semantic Reasoning, Privacy	Full (Semantic WoT with CoAP and DRDs)	High (evaluated in smart hospital use case)
[16]	Public/Permissionless (IOTA Tangle)	Privacy, Integrity, E2E Encryption, Availability	Full (MQTT/MQTT-SN + WoT)	High (validated via simulation and modular design)
[17]	Permissioned (Hyperledger Fabric)	Privacy, Confidentiality, Integrity, Authentication, Traceability	Partial (Node-RED + IoMT integration)	Medium (validated via proof of concept)

4 WoT Security Challenges

Although blockchain technologies have introduced promising solutions to enhance the security, trust, and transparency of WoT environments, several critical challenges remain unresolved as shown below. These challenges highlight the need for continued research to develop lightweight, interoperable, and resilient security frameworks tailored to the unique demands of WoT infrastructures.

- Lack of lightweight, privacy-preserving authentication mechanisms, in particular the privacy of identities, suitable for constrained WoT/IoT devices, in a major challenge which is particularly relevant in decentralized identity frameworks such as those proposed in [12]. Future research could explore the integration of zero-knowledge proofs to enable secure authentication without revealing sensitive identity attributes, even on resource-constrained hardware.
- The scalability and energy efficiency of consensus protocols, especially in real-time applications, as highlighted in [15]
- Dynamic trust management in decentralized WoT environments, including the automation of trust scoring and revocation, is still underdeveloped [12, 14, 18].
- Lack of efficient and scalable revocation mechanisms for Verifiable Credentials (VCs) tailored to IoT devices, which often operate under limited connectivity, storage, and computational power [19]
- Secure onboarding and life cycle management of devices remains complex, especially in large-scale deployments where decentralized identity and trust must be maintained without centralized control [18, 19].

Moreover, despite the progress achieved in recent years, the realization of a fully secure WoT environment remains a complex challenge, as detailed in the following list:

- **Complexity and heterogeneity of WoT environments:** WoT systems involve a wide variety of devices, protocols, standards, and use cases. This heterogeneity makes it challenging to implement unified and consistent security strategies. Furthermore, the evolving nature of WoT adds to the difficulty of maintaining up-to-date security postures that can adapt to new components and interactions [11] [15].
- **Insufficient modeling of attack scenarios:** The lack of formal or visual representations (such as UML diagrams or interaction sequences) limits the understanding of how specific attacks are carried out and which vulnerabilities they exploit. For example, modeling a MITM attack in a smart irrigation system helps identify critical points where interventions can prevent system compromise [2].
- **Lack of standardized security patterns:** Developers often lack structured, reusable solutions for common security problems in WoT. The absence of security patterns tailored to WoT environments hampers the implementation of consistent and efficient defenses aligned with known threats [20].

- **Absence of security reference architectures (SRAs):** Security reference architectures provide a structured framework to guide the design and implementation of secure systems. Their lack in WoT contexts leaves a gap in systematic threat coverage, making it difficult to align security solutions with evolving use cases such as smart cities or industrial automation [21].
- **Challenges in applying the Zero Trust model:** The Zero Trust approach, which assumes no implicit trust and requires continuous verification, is not yet widely applied in WoT systems. Research is needed to adapt its principles—such as fine-grained access control, network micro-segmentation, and continuous authentication—to the constraints of WoT environments [22].
- **Limited use of AI and machine learning for proactive security:** Although AI and ML hold potential for anomaly detection, threat prediction, and automated response, their integration into WoT remains limited. Challenges include processing heterogeneous data in real time and deploying models on resource-constrained devices without compromising performance [2].
- **Securing microservice-based WoT architectures:** The growing use of microservices improves modularity and scalability but introduces new attack surfaces. Security challenges include ensuring secure communication between services, isolating compromised components, and implementing effective monitoring and rapid incident response mechanisms [2].

5 Conclusion

The integration of blockchain and distributed ledger technologies into the WoT has led to a diverse array of architectures and frameworks aimed at enhancing security, trust, interoperability, and scalability in IoT ecosystems. From smart contract-based digital twins and decentralized identity management systems to semantic-aware resource discovery and distributed event storage, the reviewed works demonstrate significant progress in addressing core challenges such as data integrity, authentication, privacy, and decentralized coordination. Solutions like Hyperledger Fabric, Ethereum, and IOTA Tangle have each contributed unique strengths, whether through permissioned control, public verifiability, or lightweight consensus mechanisms.

Despite these advancements, many research challenges remain open. Future work must address the privacy-preserving authentication of constrained devices, the efficient integration of zero-knowledge proofs, and the optimization of consensus protocols for real-time, low-power environments. Additionally, ensuring seamless interoperability across heterogeneous WoT platforms, managing dynamic trust in decentralized identity systems, and supporting scalable, geo-distributed deployments without compromising latency or availability are critical directions. As the WoT continues to evolve alongside Web 3.0 paradigms, these challenges offer fertile ground for innovation in secure, intelligent, and autonomous IoT infrastructures.

References

1. Statista: Number of Internet of Things (IoT) connections worldwide from 2022 to 2023, with forecasts from 2024 to 2034 (2025), https://www.statista.com/statistics/1183457/iot-connected-devices-worldwide/, accessed: 27 June 2025
2. Albarrak, K.M.: Securing the future of web-enabled IoT: A critical analysis of web of things security. Applied Sciences **14**(23) (2024). https://doi.org/10.3390/app142310867, https://www.mdpi.com/2076-3417/14/23/10867
3. Nguyen, D.C., Pathirana, P.N., Ding, M., Seneviratne, A.: Integration of blockchain and cloud of things: Architecture, applications and challenges. IEEE Communications Surveys & Tutorials **22**(4), 2521–2549 (2020). https://doi.org/10.1109/COMST.2020.3020092
4. Ghorbian, M., Ghobaei-Arani, M.: A blockchain-enabled serverless security mechanism for IoT-based drones. In: Building Cybersecurity Applications with Blockchain and Smart Contracts, pp. 55–82. Springer (2024)
5. (W3C), W.W.W.C.: Web of Things (WoT) security and privacy guidelines (2020), https://www.w3.org/TR/wot-security/
6. El Jaouhari, S., Bouabdallah, A., Bonnin, J.M.: Chapter 14 - security issues of the web of things. In: Sheng, Q.Z., Qin, Y., Yao, L., Benatallah, B. (eds.) Managing the Web of Things, pp. 389–424. Morgan Kaufmann, Boston (2017). https://doi.org/10.1016/B978-0-12-809764-9.00018-4, https://www.sciencedirect.com/science/article/pii/B9780128097649000184
7. Catuogno, L., Turchi, S.: The dark side of the interconnection: Security and privacy in the web of things. In: 2015 9th International Conference on Innovative Mobile and Internet Services in Ubiquitous Computing. pp. 205–212 (2015). https://doi.org/10.1109/IMIS.2015.86
8. Aguzzi, C., Gigli, L., Zyrianoff, I., Roffia, L.: Zion: A scalable w3c web of things directory. In: 2024 IEEE 21st Consumer Communications & Networking Conference (CCNC). pp. 1–6 (2024). https://doi.org/10.1109/CCNC51664.2024.10454685
9. Pittaras, I., Polyzos, G.C.: Secure and efficient web of things digital twins using permissioned blockchains. In: 2022 7th International Conference on Smart and Sustainable Technologies (SpliTech). pp. 1–5 (2022). https://doi.org/10.23919/SpliTech55088.2022.9854219
10. Wang, S.Y., Hsu, Y.J., Hsiao, S.J.: Integrating blockchain technology for data collection and analysis in wireless sensor networks with an innovative implementation. In: 2018 International Symposium on Computer, Consumer and Control (IS3C). pp. 149–152 (2018). https://doi.org/10.1109/IS3C.2018.00045
11. Gigli, L., Sciullo, L., Montori, F., Marzani, A., Di Felice, M.: Blockchain and web of things for structural health monitoring applications: A proof of concept. In: 2022 IEEE 19th Annual Consumer Communications & Networking Conference (CCNC). pp. 699–702 (2022). https://doi.org/10.1109/CCNC49033.2022.9700679
12. Luecking, M., Fries, C., Lamberti, R., Stork, W.: Decentralized identity and trust management framework for Internet of Things. In: 2020 IEEE International Conference on Blockchain and Cryptocurrency (ICBC). pp. 1–9 (2020). https://doi.org/10.1109/ICBC48266.2020.9169411
13. Bellini, P., Branchi, E., Collini, E., Palesi, L.A.I., Nesi, P., Pantaleo, G.: Certifying entity models, entities and data messages on IoT/WoT platforms via blockchain. Computer Networks **265**, 111281 (2025). https://doi.org/10.1016/j.comnet.2025.111281, https://www.sciencedirect.com/science/article/pii/S138912862500249X
14. Durand, A., Gremaud, P., Pasquier, J.: Decentralized web of trust and authentication for the Internet of Things. IoT '17, Association for Computing Machinery, New York, NY, USA (2017). https://doi.org/10.1145/3131542.3140263
15. Zaghouani, K., Djamaa, B., Yachir, A., Mahmoudi, S.: Semchain: A blockchain-based semantic discovery on distributed resource directories for the Internet of Things. Internet of Things **31**, 101591 (2025). https://doi.org/10.1016/j.iot.2025.101591, https://www.sciencedirect.com/science/article/pii/S2542660525001040
16. Chiti, F., Gandini, G.: Distributed ledger as a service: A web 3.0-oriented architecture. Journal of Sensor and Actuator Networks **12**(4) (2023). https://doi.org/10.3390/jsan12040057, https://www.mdpi.com/2224-2708/12/4/57

17. El Jaouhari, S., Tamani, N.: An end-to-end secure solution for IoMT data exchange. In: Andreoni, M. (ed.) Applied Cryptography and Network Security Workshops. pp. 3–15. Springer Nature Switzerland, Cham (2024)
18. Putra, G.D., Putra, B.R.O.: Endorsement-driven blockchain SSI framework for dynamic IoT ecosystems (2025), https://arxiv.org/abs/2507.09859
19. Mazzocca, C., Acar, A., Uluagac, S., Montanari, R.: Evoke: efficient revocation of verifiable credentials in IoT networks. In: Proceedings of the 33rd USENIX Conference on Security Symposium. SEC '24, USENIX Association, USA (2024)
20. Sardar, R., Anees, T., Al-Shamayleh, A.S., Mehmood, E., Khalil, W., Akhunzada, A., Shaikh, F.S.: Challenges in detecting security threats in WoT: a systematic literature review. Artificial Intelligence Review **58**(7), 196 (2025)
21. Su, X., Gilman, E., Liu, X.: Towards semantic web of things: Reference architecture and gap analysis. In: 2023 IEEE International Conferences on Internet of Things (iThings) and IEEE Green Computing & Communications (GreenCom) and IEEE Cyber, Physical & Social Computing (CPSCom) and IEEE Smart Data (SmartData) and IEEE Congress on Cybermatics (Cybermatics). pp. 151–158 (2023). https://doi.org/10.1109/iThings-GreenCom-CPSCom-SmartData-Cybermatics60724.2023.00048
22. Liu, C., Tan, R., Wu, Y., Feng, Y., Jin, Z., Zhang, F., Liu, Y., Liu, Q.: Dissecting zero trust: Research landscape and its implementation in IoT. Cybersecurity **7**(1), 20 (2024)

A Smart Contract-Based Blockchain Solution for Key Revocation in IoT/WoT Environments

Sami Bettayeb, Mohamed-Lamine Messai, Sofiane Mounine Hemam, and Belal Alsinglawi

1 Introduction and Motivation

The emergence of the Internet of Things (IoT) has brought about a significant technological shift, enabling us to interact with smart homes, wearable devices, and industrial automation in ways we never thought possible. Building upon IoT, the Web of Things (WoT) extends these capabilities by standardizing interoperability between diverse connected devices, making integration, communication, and control more seamless and scalable [1]. For instance, in the industrial field, connected machines communicate with each other, optimizing production lines and facilitating predictive maintenance. WoT frameworks further enhance this by enabling unified access to heterogeneous devices through standardized web protocols, simplifying data exchange and decision-making across different platforms [2]. This translates to

Sami Bettayeb and Mohamed-Lamine Messai contributed equally to this work.

S. Bettayeb
ICOSI Laboratory, Abbes Laghrour University, Khenchela, Algeria
e-mail: sami.bettayeb@univ-khenchela.dz

M.-L. Messai (✉)
Université Lumière Lyon 2, Universite Claude Bernard Lyon 1, ERIC, Lyon, France
e-mail: mohamed-lamine.messai@univ-lyon2.fr

S. M. Hemam
Abbes Laghrour University, Khenchela, Algeria
e-mail: hemam.sofiane@univ-khenchela.dz

B. Alsinglawi
Zayed University, Abu Dhabi, UAE
e-mail: belal.alsinglawi@zu.ac.ae

S. El Jaouhari, N. El Madhoun (eds.), *Blockchain-based Web of Things Security: Advances, Challenges, and Solutions*, Signals and Communication Technology,
https://doi.org/10.1007/978-3-032-07671-7_4

increased efficiency, reduced downtime, and improved safety for workers. However, security and privacy concerns have become significant challenges as the number of connected devices increases. One of the most critical security issues in both IoT and WoT networks is key management on devices with limited resources, as ensuring secure authentication and communication remains a fundamental challenge in highly interconnected environments.

Robust key management solutions are very important, especially in the dynamic and interconnected networks such as IoT networks, and present significant challenges [3]. The IoT devices often have limited processing power, memory, and energy resources, making it difficult to implement traditional key management schemes. Robust key management serves as a crucial pillar for ensuring the integrity and confidentiality of data exchanged within IoT networks. In this context, effective key management involves a meticulous process of generating, distributing, managing, and revoking cryptographic keys. These keys play a pivotal role in encrypting and decrypting communication, ensuring that sensitive information remains secure from unauthorized access. Beyond its cryptographic function, key management also contributes to secure device authentication and facilitates the revocation and update of keys as needed.

Key revocation is an essential phase of key management schemes. A compromised key without revocation, it remains a vulnerability, potentially allowing unauthorized access to sensitive information. Revocation acts as a shield, effectively deactivating compromised keys. However, traditional systems require a central authority to manage and distribute keys, creating single points of failure vulnerabilities and significant communication overhead. The communication overhead metric for the key revocation step in a key management solution can be defined as the total amount of data transmitted between devices or entities involved in the key revocation process. This includes the exchanged messages for revocation and acknowledges the receipt of revocation information. Key management solutions with a central authority lead to potential security breaches and high energy consumption because of the high number of exchanged messages between a central authority and the IoT devices. To address these challenges, blockchain technology has emerged as a potential solution. Blockchain's properties, such as decentralization, immutability, and security, make it ideal for various applications, including key management and revocation in IoT networks. Instead of relying on a central authority to maintain and distribute revocation lists, blockchain-based solutions store this information on a distributed network of gateways which reduces the number of exchanged messages and consequently the amount of consumed energy. This chapter proposes a new solution combining blockchain for key revocation using smart contracts in IoT networks. This solution uses blockchain to decentralize the key revocation process and enhance security in IoT networks. Using blockchain and smart contracts for the key revocation phase in IoT networks offers several advantages [4]. Blockchain's decentralized nature ensures that key revocation data is tamper-proof and securely recorded across multiple gateways, eliminating single points of failure, eliminating the need for individual IoT devices to maintain a complete revocation list, saving storage space. Smart contracts allow automated key revocation process, enabling instant and transparent execution of revocation without human intervention. This automation ensures that revoked keys are swiftly removed. Furthermore, the transparency of blockchain provides an immutable audit trail, allowing stakeholders to verify key revocations and ensuring accountability. This combination of enhanced security, automation, and transparency makes blockchain and smart contracts an ideal solution for managing key revocation in IoT networks, where the scale demands robust and reliable mechanisms. Our solution demonstrates the feasibility and effectiveness of using blockchain-based smart contracts for key revocation in devices with limited resources.

This chapter aims to analyze the potential benefits of blockchain-based key revocation in IoT networks, including improved security, reduced communication overhead, and energy consumption. A first version of this work has been published in [5], and in this chapter, we extend it with more comparisons and evaluation.

The proposed solution has the following features:

- **Decentralization**: The proposed solution is decentralized, meaning there is no central authority to manage the key revocation process. Instead, the blockchain network participants are responsible for the key revocation process.
- **Transparency**: The proposed solution is transparent, meaning that all the participants can see each device's key revocation process and revocation status.
- **Reduced communication overhead and energy consumption**: The proposed solution reduces the communication overhead by using blockchain-based smart contracts to manage the key revocation process.
- **Security**: The proposed solution is secure because it uses blockchain technology to manage the key revocation process, making it difficult for an attacker to compromise the system.

By incorporating blockchain into our key revocation solution, this proactive approach minimizes the window of opportunity for potential security breaches, ensuring enhanced security and system transparency. Furthermore, the proposed approach significantly reduces communication overhead and energy consumption, making it suitable for resource-constrained IoT devices.

The chapter's organization is as follows: Sect. 2 provides a brief background on Blockchain, Sect. 3 comprehensively reviews related work in key management and revocation in IoT networks, Sect. 4 describes the proposed blockchain-based key revocation mechanism, including the smart contract design and implementation, Sect. 5 evaluates the performance and security of the proposed mechanism and compares it with existing approaches, and finally, Sect. 6 concludes and highlights future research directions in this field.

2 Preliminaries

2.1 Blockchain

Blockchain is a decentralized, distributed ledger technology that records transactions in a secure and transparent manner. It is a growing field of research with applications in various domains, including finance, supply chain management, and healthcare. It was used in 2008 by Nakamoto in a paper on decentralized peer-to-peer digital currency [6]. The primary goal of Blockchain is to create a shared ledger distributed across all parties in the network, which contains all transactions made in the network and is synchronized using a consensus algorithm. Blockchain offers several benefits: distribution, security, traceability, and immutability [7, 8]. Different consensus algorithms have been proposed to ensure a common consensus among participants, such as proof of work [6], proof of stake [9], delegated proof of stake [10], and practical Byzantine fault tolerance [11]. Proof of work is the first algorithm introduced in Blockchain as a consensus, which is used in Bitcoin [6]. It is a computationally intensive algorithm that requires much computing power to solve a cryptographic puzzle. Proof of stake is another consensus algorithm that does not require much computing power. Instead, participants must stake a certain amount of virtual currency to participate in the consensus algorithm. Delegated proof of stake is a variation of proof of stake that requires participants to vote for a certain number of delegates to participate in the consensus algorithm. Practical Byzantine fault tolerance is a consensus algorithm that requires a certain number of participants to agree on a transaction before it is added to the Blockchain. Blockchain technology comes in various types, each designed to serve specific purposes and address different use cases.

2.2 Types of Blockchain

Blockchain can be classified into three types: public, private, and consortium. A public blockchain lets anyone join the network and participate in the consensus algorithm. In contrast, only a restricted number of participants from the same organization can participate in private Blockchain. In the Blockchain consortium, participants from various organizations can participate in the consensus algorithm.

In 1994, Nick Szabo first introduced smart contracts when he proposed a digital contract [12]. They are self-executing programs on Blockchain, are utilized to automate contract execution between parties, and reduce time and cost without needing a trusted third party. They are deployed on the Blockchain and are executed when a specific condition is met. They are immutable and cannot be modified once deployed on the Blockchain.

Blockchain has numerous applications in various fields, including key management and data integrity. Several studies have utilized Blockchain technology, such as those presented in [7, 13–17].

2.3 *Smart Contract*

In IoT networks, smart contracts can play a pivotal role in streamlining the key revocation process within a key management solution. A smart contract is a self-executing program that automatically enforces rules and conditions on a blockchain without the need for intermediaries. For key revocation, a smart contract can be designed to manage the revocation process in a secure, automated, and decentralized manner.

In our solution, a key revocation is needed in three cases, i.e., a node compromising, a leaving node, and drained node. For instance, when an attacker captures an IoT device node, the key of that node is compromised. When the compromised node is detected (we suppose that there is a mechanism to detect compromised nodes, such as intrusion detection system), the server submits a revocation request to the smart contract deployed on the blockchain. The smart contract verifies the authenticity of the request and records the revocation event on the blockchain. This information is then accessible to all nodes in the IoT network, enabling IoT devices to efficiently query the blockchain to check the revocation status of keys without relying on a centralized authority.

Using smart contracts for key revocation offers several advantages:

- Automation: The entire process is automated, reducing the need for human intervention or additional infrastructure for managing revocations.
- Immutability : The blockchain ensures that revocation records cannot be tampered with, providing a secure and reliable log of revocation events.
- Decentralization: Smart contracts eliminate the need for a central authority to manage key revocations, reducing single points of failure and improving scalability for IoT networks.

3 Related Work

Multiple research papers introduced different approaches and solutions to the key management problem with its different phases: key generation, distribution, pairwise key establishment, group-wise key establishment, new node addition, key refresh, and revocation. In this section, we discuss the most relevant works in key revocation.

Nafi et al. [18] proposed a matrix-based key management scheme, on which the key revocation is one of its phases. They distinguish between two scenarios: the first one is when a node gets compromised and detected by the gateway, and the second one is when a node needs to leave the network voluntarily. In the first scenario, upon detecting malicious activity from a node by a gateway, the gateway starts revocation by deleting the node's keys from its memory, its secrets from its matrix, and its identifier from its neighboring vector. It then broadcasts a revocation message to all nodes in the network. Upon receiving the revocation message, each node, as the gateway already did, deletes the compromised node's keys from its memory, its secrets from its matrix, and the malicious node identifier from the neighbors' vector. It then replies to the gateway with an acknowledgment message confirming that it has successfully revoked the compromised node. The gateway then starts the rekeying process to establish new keys with the remaining nodes in the network. The remaining nodes also start rekeying to establish new keys with their neighbors. In the second scenario, when a node needs to leave the network voluntarily, it sends a leave message to its neighbors. Upon receiving the leave message, each node deletes the leaving node's keys from its memory, its secrets from its matrix, and the leaving node identifier from the neighbors' vector. It then replies to the leaving node with an acknowledgment message confirming that it has successfully revoked the leaving node. The leaving node leaves the network after receiving the acknowledgment message from all its neighbors and erasing its memory. The remaining nodes start rekeying to establish new keys with their neighbors.

Nafi et al. in another work presented in [19] introduced a node deletion phase that involves neighboring nodes of a compromised or leaving node deleting it from their neighboring tables and adjusting their polynomials by reducing them by one term. Additionally, they remove the shared pairwise keys from their memory. Nonetheless, in their approach, the authors did not provide details regarding detecting a revoked node and the communications necessary for handling such an event.

As stated by the authors in [20], neighboring nodes are responsible for the key revocation process of a node. No messages are sent to neighbor nodes, yet each node invalidates the key of any node that fails to send a message within a predetermined time interval. Additionally, nodes perform a key generation phase every T_r, and the server transmits a set of new one-to-one functions to all nodes every $(2 * T_r)$. If a node compromising attack is detected, the server sends a refresh message to all nodes via the gateway node. This message includes the number of attacks to refresh the value of T_r. By receiving and responding to the server's refresh message, the communication overhead of this solution remains comparable to that of the approach proposed in [18].

In their study, Mesmoudi et al. [21] delineate two primary scenarios for key revocation within their scheme as their network model includes two distinct types of nodes: the cluster head (CH) and the cluster member (CM). The first scenario occurs when the base station (BS) identifies a compromised CH. At this point, it revokes the compromised CH's key and associated CMs, initiating a key refresh for the remaining nodes. The second scenario arises when a CH detects a compromised CM, prompting it to revoke the key of the affected CM and notify the other CMs within the cluster while initiating the key renewal process. Notably, the scheme does not account for key revocation when a node leaves the network or experiences battery depletion.

Kandi et al. in [14] proposed a decentralized blockchain-based scheme that featured two layers—one for key management and the other for nodes' key management. To revoke the key of a leaving node, they sent a refresh message to the node set of the leaving node containing the identifier and a randomly generated K_r that was used to refresh the keys of the remaining nodes. They also broadcast a refresh message to other sets of nodes containing the identifier of the leaving node set and K_r. When the second layer of this scheme, composed of Blockchain participants (BPs), received information about the leaving node, it created a new transaction containing the identifier and the other information about the leaving node. Finally, all the BPs verify this transaction by checking the signature, hashes, and validity. This scheme decentralizes the key revocation process and resolves the single point of failure problem faced by centralized schemes. However, the constrained nodes' participation in the key revocation process produces a high communication overhead.

Recently in [16], a blockchain-based key management solution is presented for distributed IoT architectures. The network is modeled as three layers, i.e., an equipment layer, an intermediate layer, and a cloud layer. The authors use hash chains and a key distribution scheme based on network security coding to secure communications. The one-way hash chains offer the identity authentication process between the intermediate layer and the equipment layer, as well as the mutual identity authentication process during device communication. The latter employs secure network coding principles to fortify the security of transmitting sensitive information. Sensors and IoT devices are in the equipment layer, and the blockchain structure is stored in the intermediate layer to avoid heavy computational to the resource-limited devices. The intermediate layer is connected to the cloud layer, which manages each blockchain network. Adding an intermediate layer in the network architecture to reduce the computing time and overhead of IoT devices makes this solution scalable. However, the IoT devices have to run heavy operations within an asymmetric cryptosystem. In addition, key revocation phase is not presented in the solution.

4 Proposed Solution

4.1 Network Model and Notations

The network model consists of three main parts: IoT devices, gateway nodes, and remote server:

- **IoT devices**: These are the primary components of the network that collect and transmit data from the environment to the gateway nodes. Ranging from small sensor nodes to large industrial sensor-equipped machines and appliances, these devices are often constrained in terms of resources such as memory, processing, power, and communication.

- **Gateway nodes**: Acting as intermediaries between the IoT devices and the remote server, these resource-rich devices collect data from the IoT devices and forward it to the remote server. Each gateway node is a participant in the blockchain network. Each pair of gateways can communicate directly or through the server.
- **Remote server**: This component is responsible for receiving data from the gateway nodes. It processes the data further, such as storing it in a database, analyzing it, visualizing it, and making it available via an API.

Each IoT device performs key establishment with each neighboring node and collects data related to the application. The gateway nodes integrate blockchain technology and manage data transmission from and to the remote server.

In this setup, each gateway is linked to the remote server and has one or more IoT devices connected to it. At the same time, each IoT device is linked to at least one gateway node, which is responsible for managing the device's keys.

The network model is illustrated in the part (a) in Fig. 1. The part (b) presents the blockchain deployed in the three gateways.

The used notations in this chapter are shown in Table 1.

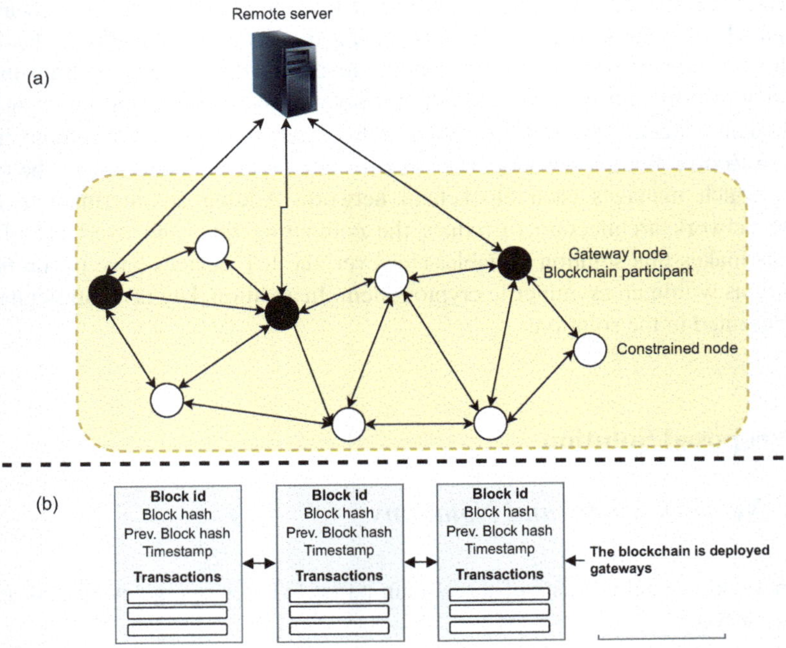

Fig. 1 Network model

Table 1 Notations

Notation	Meaning
l_{ms}	The sent message size in bytes
l_{mr}	The received message size in bytes
d	The total number of a node's neighbors
n	The total number of nodes in the network
n_s	The number of set member nodes
n_{gm}	The number of member nodes that are linked to the gateway g
n_c	The number of cluster members
$EPSB$	The consumed energy per sent byte
$EPRB$	The consumed energy per received byte

4.2 Assumptions

It is assumed that:

- Bidirectional wireless communication: IoT devices in the network can both send and receive messages to and from their neighbors.
- The IoT devices have limited processing capacity, memory, and energy resources (battery-operated). The IoT devices may join or leave the network.
- The proposed scheme is built based on a key management scheme that leverages blockchain technology to at least administer its key distribution phase.
- The utilized consensus algorithm is of the proof-of-stake (PoS) variety, owing to its heightened energy efficiency when compared to the proof-of-work (PoW) consensus [22].
- The used blockchain type is private, as it is deemed more efficient in energy consumption and transaction processing time when compared to the public blockchain.
- The used signature scheme is the elliptic curve digital signature algorithm (ECDSA) [23], owing to its widespread adoption and its use in the Bitcoin blockchain [24].

4.3 Solution Steps

A node can be revoked in the proposed solution under three distinct scenarios: firstly, as a compromised node. Secondly, when the node voluntarily exits the network, and thirdly, when the node's battery power is completely drained. We detail each case in the following subsections.

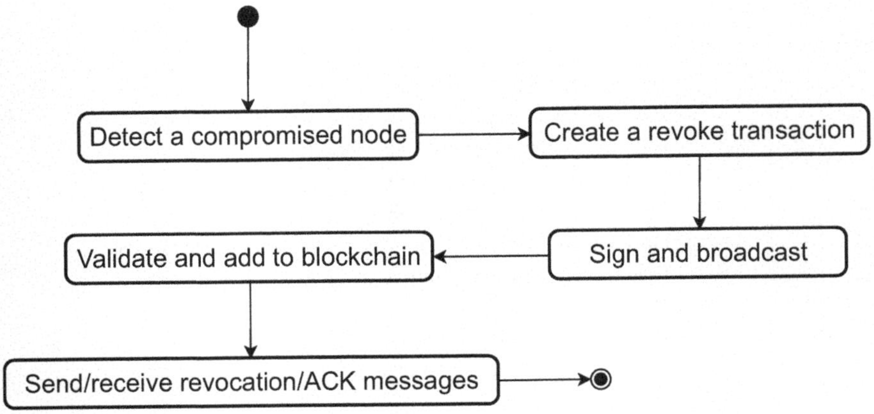

Fig. 2 Flowchart of compromised node revocation process

4.3.1 Compromised Node

A compromised node is one that an adversary has compromised. In such case, the node becomes a security threat to the network, necessitating the revocation of all its sensitive data, including cryptographic secrets and keys. The proposed solution enables the revocation of a compromised node through the blockchain and key management layers. The process is initiated by a gateway, a blockchain participant, who analyzes the node's behavior on the blockchain ledger to detect malicious activity. It creates a revocation transaction containing the node's information, including its identifier and revocation time, and the reason for revocation. This transaction is then signed and broadcast to the blockchain network. Each blockchain participant that receives the transaction validates it by executing the appropriate smart contract, checking if the node already belongs to the network. If the node is not found, the transaction is rejected. If malicious activity is detected, the transaction is accepted and added to the blockchain ledger. Once added, each blockchain participant checks for any nodes linked to the compromised node and broadcasts a revocation message to them, instructing them to erase any cryptographic keys of the revoked node. Linked nodes reply with an acknowledgment message upon completion of this task. If the node is not found to be malicious, the transaction is rejected and not added to the blockchain ledger.

The compromised node revocation process is illustrated in Fig. 2.

4.3.2 Leaving Node

A leaving node is one that voluntarily exits the network, perhaps to conserve battery power. Although not a security threat, the node's cryptographic keys and sensitive data must still be revoked to prevent future attacks. The leaving node initiates

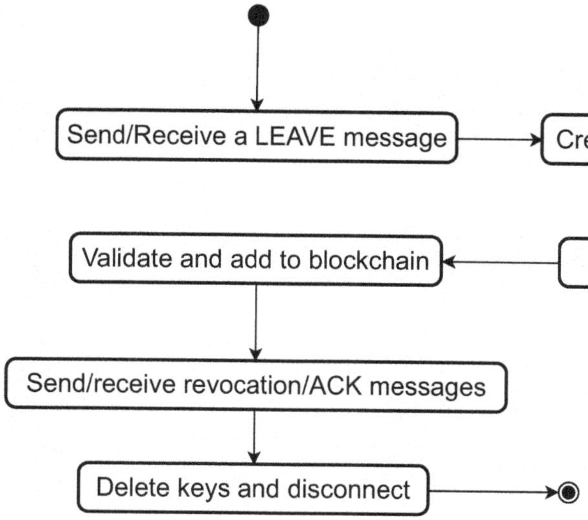

Fig. 3 Flowchart of leaving node revocation process

this process by sending a revocation message to its linked gateway. The gateway then creates, signs, and broadcasts a revocation transaction containing the node's information to the blockchain network. Each blockchain participant validates this transaction by executing the associated smart contract. If the node is not found in the network, the transaction is rejected. If found, the transaction is added to the blockchain ledger, and each blockchain participant checks for any nodes linked to the leaving node, instructing them to erase its cryptographic keys. Acknowledgment messages are exchanged upon completion of these steps.

The leaving node revocation process is illustrated in Fig. 3.

4.3.3 Drained Node

A drained node is one whose battery power is completely depleted. Similar to a leaving node, it is not considered a security threat, but its cryptographic keys and sensitive data must be revoked. The revocation process is initiated by a blockchain participant who detects that the node's battery is drained by analyzing its last activity. If the node is confirmed as drained, the blockchain participant creates a revocation transaction containing the node's information and broadcasts it to the network. Each blockchain participant validates the transaction by checking specific conditions through a smart contract. If any condition is not met, the transaction is rejected. Otherwise, it is added to the blockchain ledger, and each blockchain participant checks for any nodes linked to the drained node, instructing them to erase its cryptographic keys. Acknowledgment messages are then exchanged to confirm the completion of these steps.

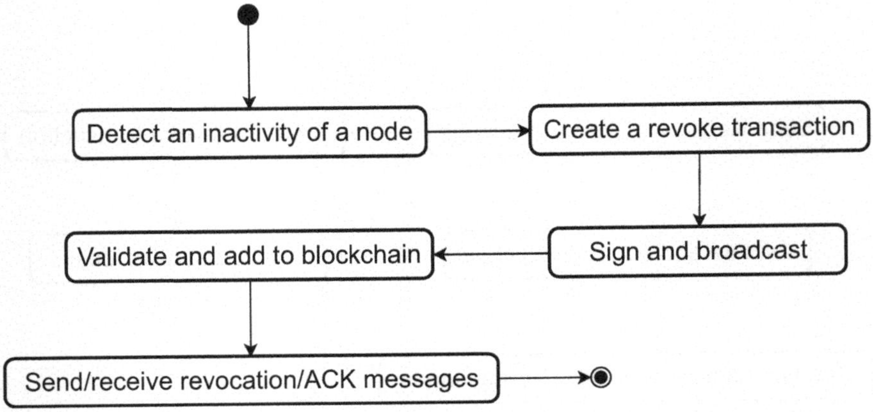

Fig. 4 Flowchart of drained node revocation process

The drained node revocation process is illustrated in Fig. 4.

5 Evaluation

In this section, we analyze the security of our proposed solution. We first discuss the security of the exchanged messages between the different constrained devices, and we discuss the security of the data stored on the blockchain ledger. Then, we compare our solution with other key revocation schemes and present simulation results.

5.1 Security Analysis

Our proposed solution is assumed to be built on a key management scheme with at least a blockchain-based key agreement phase. It uses its different cryptographic primitives to ensure the security of the exchanged messages between the different constrained devices.

The transparency and immutability inherent to blockchain technology play a vital role in our solution. All transactions related to key revocations are recorded permanently and transparently on the blockchain ledger. This allows for complete auditability and prevents any unauthorized changes to the revocation status of keys. Furthermore, smart contracts deployed on the blockchain automate key management processes. These self-executing contracts enforce predefined rules and procedures for key revocation, eliminating the need for manual intervention and minimizing the risk of human error. Furthermore, the blockchain ledger does not store sensitive materials like cryptographic secrets and keys. It only stores the different transactions signed by the different blockchain participants.

5.2 *Performance Analysis*

In this section, we present the performance evaluation of our proposed solution. We first present a comparative study of different key revocation schemes. Then, we present the performance evaluation of our proposed solution through experiments and simulations regarding communication, computation overheads, and energy consumption.

5.2.1 Comparative Study of Different Key Revocation Schemes

Tables 2, 3, and 4 present a comparative study regarding the communication overhead and energy consumption of our proposed solution with some of the existing ones, namely, [18, 20, 21], and [14]. The tables present the communication overhead for each node type as the size of the sent and the received bytes during the key revocation process. The energy consumption is calculated as the number of bytes multiplied by each node's energy consumption per byte. Moreover, the

Table 2 Comparison of our solution with some existing solutions for compromised node detection

Solution	Node type	Communication	Energy consumption	Nodes count	Total energy consumption
[21]	CH	$l_{ms} + (n_c - 1) * l_{mr}$	$EPSB * l_{ms} + EPRB * (n_c - 1)l_{mr}$	1	$EPSB * l_{ms} + EPRB * (n_c - 1)l_{mr}$
	CM	$l_{ms} + l_{mr}$	$EPSB * l_{ms} + EPRB * l_{mr}$	n_c	$n_c * (EPSB * l_{ms} + EPRB * l_{mr})$
[18, 20]	Gateway	$l_{ms} + n * l_{mr}$	$EPSB * l_{ms} + n * EPRB * l_{mr}$	1	$EPSB * l_{ms} + n * EPRB * l_{mr}$
	Constrained	$l_{ms} + l_{mr}$	$EPSB * l_{ms} + EPRB * l_{mr}$	n	$n * (EPSB * l_{ms} + EPRB * l_{mr})$
Our solution	Gateway	$l_{ms} + n * l_{mr}$	$EPSB * l_{ms} + n * EPRB * l_{mr}$	1	$EPSB * l_{ms} + n * EPRB * l_{mr}$
	Constrained	$l_{ms} + l_{mr}$	$EPSB * l_{ms} + EPRB * l_{mr}$	n_{gm}	$n_{gm} * (EPSB * l_{ms} + EPRB * l_{mr})$

Table 3 Comparison of our solution with some of the existing solutions in the case of a leaving node

Solution	Node type	Communication	Energy consumption	Nodes count	Total energy consumption
[18, 20]	Leaving	$l_{ms} + d * l_{mr}$	$EPSB * l_{ms} + EPRB * d * l_{mr}$	1	$EPSB * l_{ms} + EPRB * d * l_{mr}$
	Others	$l_{ms} + l_{mr}$	$EPSB * l_{ms} + EPRB * l_{mr}$	d	$d * (EPSB * l_{ms} + EPRB * l_{mr})$
[14]	Leaving	$(n_s - 1)l_{ms} + (n - 1)l_{mr}$	$(n_s - 1)l_{ms} * EPSB + (n - 1)l_{mr} * EPRB$	1	$(n_s - 1)l_{ms} * EPSB + (n - 1)l_{mr} * EPRB$
	Others	$l_{ms} + l_{mr}$	$EPSB * l_{ms} + EPRB * l_{mr}$	n	$n * (EPSB * l_{ms} + EPRB * l_{mr})$
Our solution	Gateway	$l_{ms} + (d + 1)l_{mr}$	$EPSB * l_{ms} + EPRB * (d + 1)l_{mr}$	1	$EPSB * l_{ms} + EPRB * (d + 1)l_{mr}$
	Others	$l_{ms} + l_{mr}$	$EPSB * l_{ms} + EPRB * l_{mr}$	n_{gm}	$n_{gm} * (EPSB * l_{ms} + EPRB * l_{mr})$

total energy consumption is calculated as the energy consumption of each node multiplied by the number of nodes of that type.

In Table 2, we compare our solution with [18, 20, 21] in the case of a compromised node. As we can see, in the case of compromising a CH in [21], it needs to send one and receive $n_c - 1$ messages by gateway nodes and needs to send one and receive one message for the constrained node. While in [18, 20], the gateway node needs to send one and receive n messages by gateway nodes and needs to send one and receive one message as [21]. In contrast, in our solution, the gateway node needs to send one and receive n_{gm} messages by the gateway nodes, and send one and receive one message as in [21] and [18, 20]. The number of the involved gateway nodes in the revocation process is the same for all the solutions and higher in the case of [18, 20] in the case of constrained nodes; this makes our solution outperform [18, 20] and has the same performance as [21] in the case of a compromised node for the gateway nodes.

In Table 3, we compare our solution with [18, 20] and [14] in the case of a leaving node. As we can see, in the case of a leaving node in [18, 20], it needs to send one and receive d messages by gateway nodes and needs to send one and receive one

Table 4 Comparison of our solution with some of the existing solutions in the case of a drained node

Solution	Node type	Communication	Energy consumption	Nodes count	Total energy consumption
[18, 20]	Gateway	$l_{ms} + (n-1)l_{mr}$	$EPSB * l_{ms} + EPRB * (n-1) * l_{mr}$	1	$EPSB * l_{ms} + EPRB * (n-1) * l_{mr}$
	Constrained	$l_{ms} + (n-1)l_{mr}$	$EPSB * l_{ms} + EPRB * l_{mr}$	n	$n * (EPSB * l_{ms} + EPRB * l_{mr})$
Our solution	Gateway	$l_{ms} + (n_{gm})l_{mr}$	$EPSB * l_{ms} + EPRB * (n_{gm})l_{mr}$	1	$EPSB * l_{ms} + EPRB * (n_{gm})l_{mr}$
	Constrained	$l_{ms} + l_{mr}$	$EPSB * l_{ms} + EPRB * l_{mr}$	n_{gm}	$n_{gm} * (EPSB * l_{ms} + EPRB * l_{mr})$

message for the constrained node. While in [14], it needs to send $(n_s - 1)$ and receive $(n - 1)$ messages by the leaving node and needs to send one and receive one message for the other nodes' types. In contrast, the gateway node must send one in our solution and receive $(d + 1)$ messages. However, the number of the involved constrained nodes in the revocation process is the smallest in our solution; this makes our solution outperform [18, 20] and [14] in the case of a leaving node for the constrained nodes.

In Table 4, we compare our solution with [18, 20] in the case of a drained node. As we can see, in the case of a drained node in [18, 20], it needs to send one and receive $(n - 1)$ messages by gateway nodes and needs to send one and receive $n - 1$ messages for the constrained node. While in our solution, the gateway node needs to send one and receive (n_{gm}) messages, and the constrained node needs to send one and receive one message. Furthermore, the number of constrained nodes involved in the revocation process is the smallest in our solution and the same for the gateway nodes; this makes our solution outperform [18, 20] in the case of a drained node for the constrained and gateway nodes.

5.2.2 Experimental Results

This section showcases the outcomes of our proposed solution through simulation. We utilized Rust programming language to implement the simulation for our solution and the two other ones [18] and [20]. We considered the network nodes' total communication overhead and energy consumption. The simulation was conducted with 100 randomly deployed nodes. The rate of the gateway nodes is 10% of the total network nodes number. The number of neighbors of each node varies between 10 to 15. We ran the simulation 1000 times for the three scenarios: compromised, leaving, and drained nodes. The energy consumption model used in this simulation is based on [25], which considers the EPSB and EPRB factors as follows: $EPSB = 59.2\,\mu J$ and $EPRB = 28.6\,\mu J$ for the transmission and reception of one byte, respectively.

Communication Overhead

Figures 5, 6, and 7 demonstrate the communication overhead required by all the limited-resource nodes in the network in the three cases of compromised, left, and drained nodes, respectively.

The efficiency of our solution surpasses that of current solutions in terms of communication overhead, as evidenced by the results; this is due to the decentralization of Blockchain and the delegation of most communication tasks to the gateway nodes. Each gateway is responsible for communicating only with its attached constrained nodes, unlike previous studies such as [18] and [20]. Furthermore, most communication is carried out by the constrained nodes, with the gateway communicating with all network nodes, thereby increasing the communication overhead of the constrained devices. In our proposal, in the event of a leaving node,

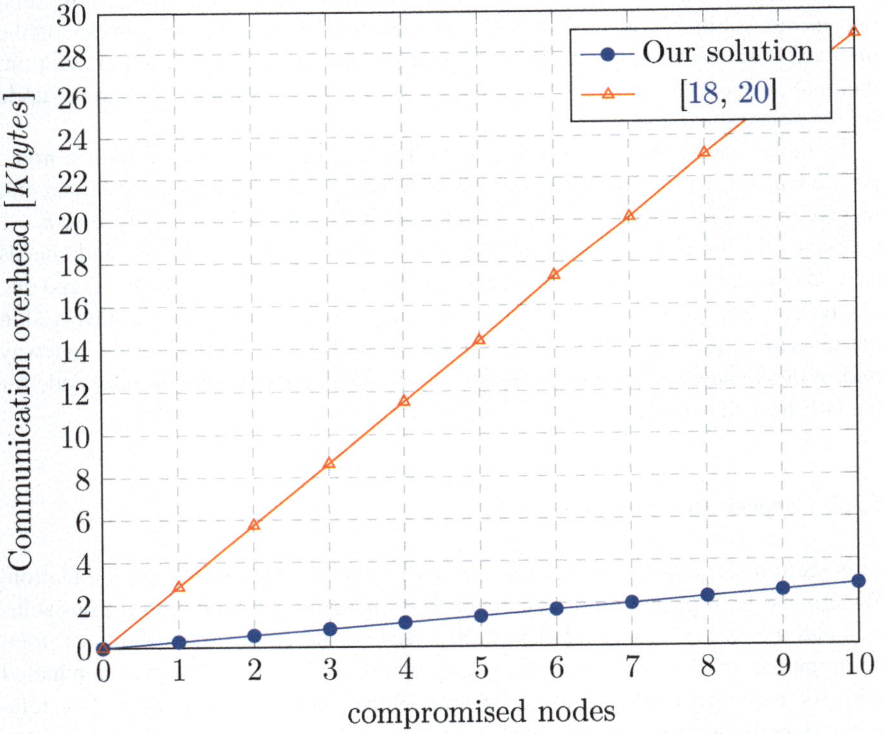

Fig. 5 Communication overhead required by all the constrained nodes in the network in the case of compromised nodes

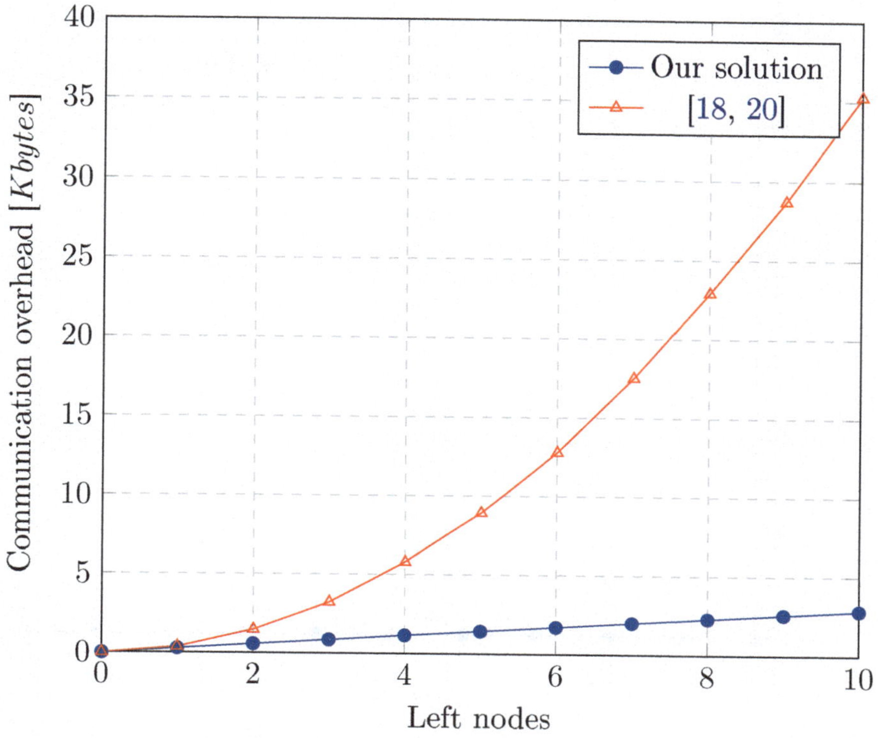

Fig. 6 Communication overhead required by all the constrained nodes in the network in the case of left nodes

the node communicates only with its gateway, which communicates with the other nodes to notify them, resulting in the sending and receiving of only one message by each of them. In contrast, in [18] and [20], the leaving node must send and receive *d* ACK messages, increasing communication overhead. In the case of a drained node, constrained IoT devices receive one message from the gateway node for the revocation process. This message is typically very small, incurring minimal communication overhead in kilobytes. In Fig. 7, the overhead is not zero, and it is optimized to minimize the impact on resource-constrained devices.

Energy Consumption

The evaluation of energy consumption in IoT networks stands as an important aspect in the effectiveness and the feasibility of security solutions.

Moreover, IoT networks often comprise resource-constrained devices with limited processing power and battery life. Optimizing energy consumption can pave the way for more efficient and durable IoT Networks.

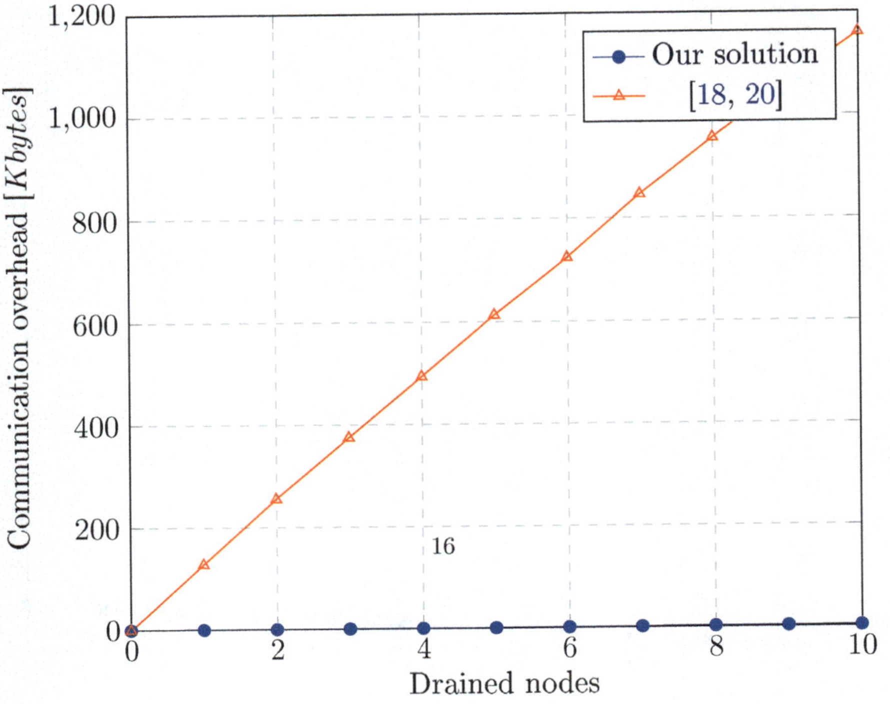

Fig. 7 Communication overhead required by all the constrained nodes in the network in the case of drained nodes

The results of the simulation on the consumed energy by the constrained nodes in the three different scenarios are presented in Figs. 8, 9, and 10. The number of affected nodes ranges from 1 to 10. As we see clearly in the figures, our approach is more efficient than existing solutions regarding consumed energy by the constrained nodes in all scenarios; this is due to the involvement of fewer constrained nodes in communication and delegating communication to gateway nodes. It is worth noting that Figs. 8, 9, and 10 share similarities with Figs. 5, 6, and 7, as the communication energy consumption is directly related to the number and the size of exchanged messages.

The energy consumption associated with key revocation of drained nodes is also very small. One message is needed for this, which requires only a negligible amount of energy. As a result, the impact of the revocation process on the battery life of IoT devices is minimal as illustrated in Fig. 10 (energy consumption is extremely low, although not entirely zero).

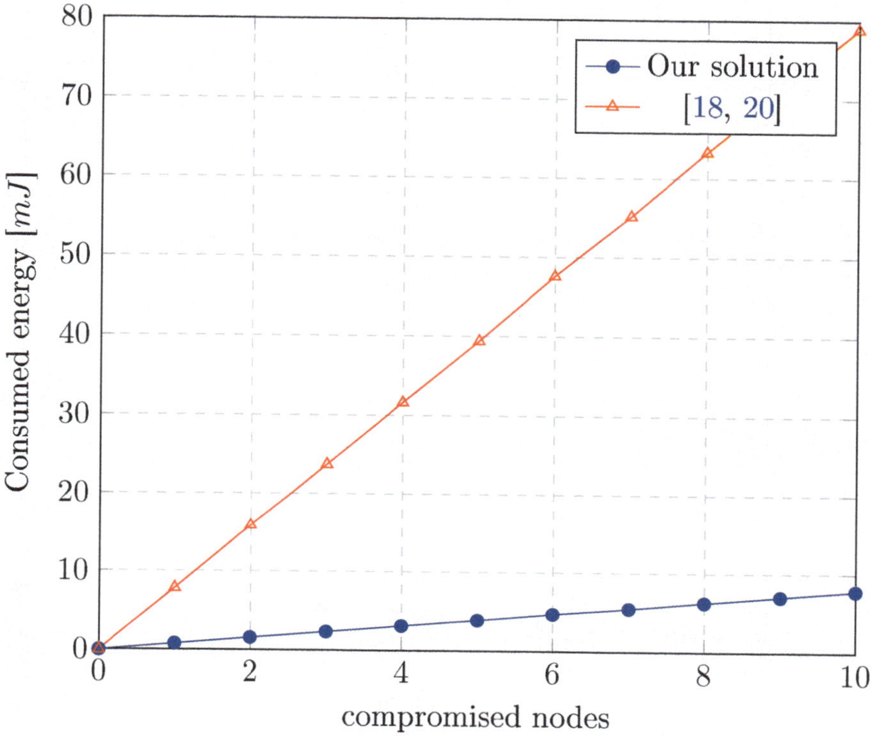

Fig. 8 Energy consumption of communication between all the constrained nodes in the network in the case of compromised nodes

These simulation results substantiate the robustness of our proposed solution and present its potential to significantly enhance energy efficiency in real-world deployment scenarios.

6 Conclusion

The issue of key revocation is crucial in any key management solution. This work proposes a blockchain and smart contract-based approach to tackle the key revocation problem in IoT networks. The proposed solution benefits from blockchain technology: transparency and automation features, which increases the security and reduces limitations like compromising the blockchain participant. In addition, it decentralizes and reduces communication overhead and energy consumption. Furthermore, security analysis and performance evaluation demonstrate that the security of the proposed solution is proportional to that of the underlying scheme. Moreover, the solution does not store any key materials in the blockchain.

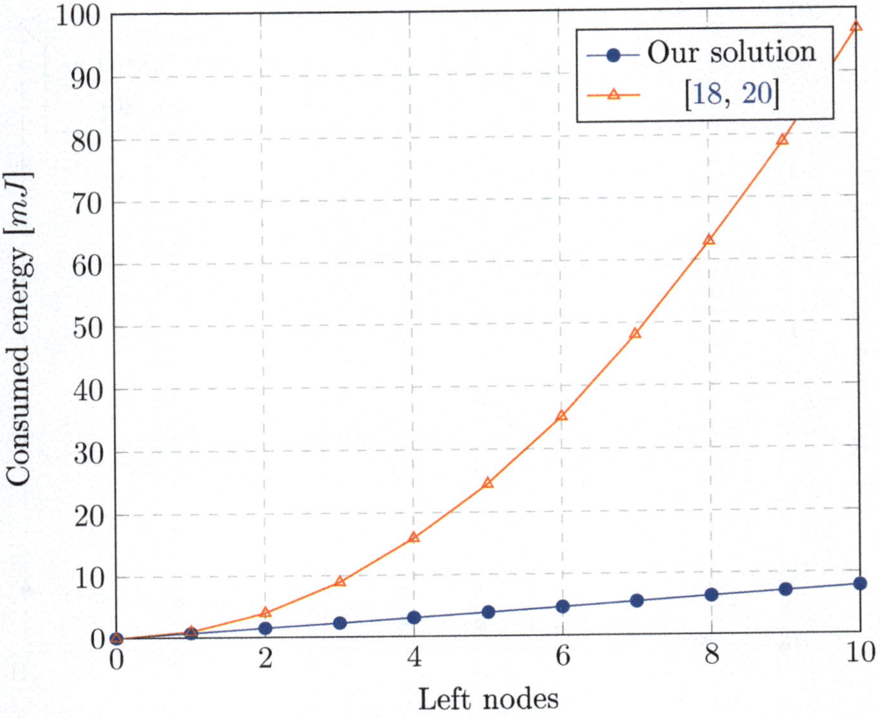

Fig. 9 Energy consumption of communication between all the constrained nodes in the network in the case of left nodes

Simulation results indicate that our proposal outperforms existing ones regarding efficiency. However, it is essential to acknowledge that the security of the proposed solution is directly related to the security of the base scheme.

We plan to extend the proposed solution in future work to include other phases of key management, such as key distribution and node addition, and to tackle their challenges. Additionally, we aim to investigate the scalability of our solution as the number of devices and revocation events within the network increases. Furthermore, we plan to explore the applicability of our solution in a WoT context, ensuring seamless and secure key management across heterogeneous and interoperable connected devices.

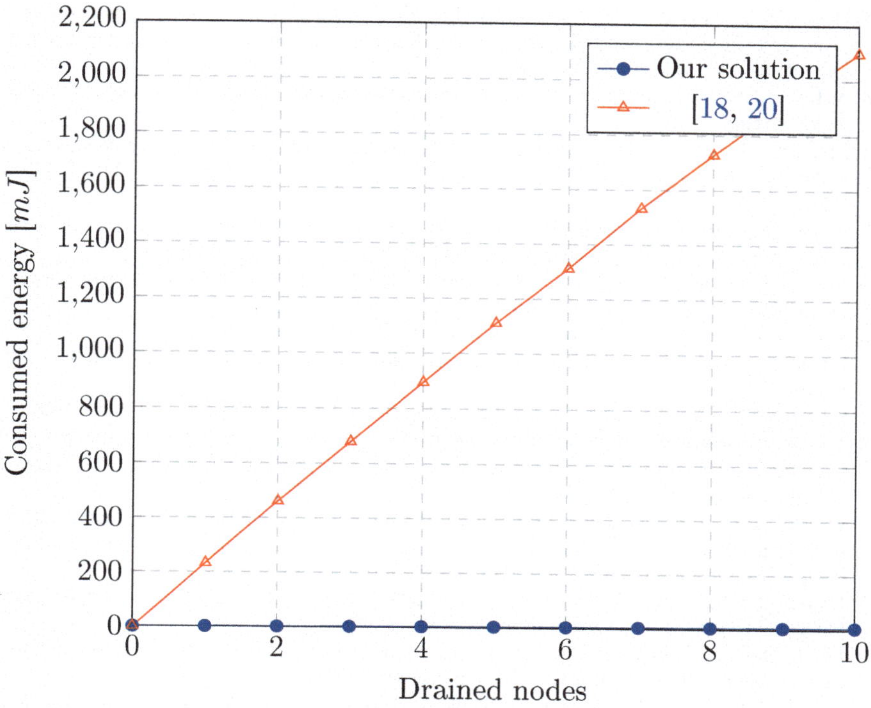

Fig. 10 Energy consumption of communication between all the constrained nodes in the network in the case of drained nodes

Declarations

- Ethics approval: Not applicable.
- Funding: No.
- Availability of data and materials: The code is available on Github: https://github.com/SAMIBETTAYEB/iot-metrics-simulation.

References

1. Manta-Caro, C., Caputo, A., Fernández-Luna, J.M.: Information retrieval for IoT and WoT: State-of-the-art, taxonomy framework and evolutionary directions. IEEE Internet of Things Journal (2024)
2. Li, B., Fan, H., Gao, Y., Dong, W.: WaWoT: Towards flexible and efficient web of things services via web assembly on resource-constrained IoT devices. IEEE Transactions on Computers (2024)
3. Messai, M.-L., Seba, H.: A survey of key management schemes in multi-phase wireless sensor networks. Computer Networks **105**, 60–74 (2016)
4. Rashid, A., Masood, A., Khan, A.u.R.: ACS-IoT: Smart contract and blockchain assisted framework for access control systems in IoT enterprise environment. Wireless Personal Communications, 1–22 (2024)

5. Bettayeb, S., Messai, M.-L., Hemam, S.M.: BKRSC-IoT: Blockchain-based key revocation using smart contracts for IoT networks. In: International Conference on Management of Digital, pp. 331–344 (2023). Springer
6. Nakamoto, S.: Bitcoin: A peer-to-peer electronic cash system. Decentralized business review, 21260 (2008)
7. Ma, M., Shi, G., Li, F.: Privacy-oriented blockchain-based distributed key management architecture for hierarchical access control in the IoT scenario. IEEE access **7**, 34045–34059 (2019)
8. Viriyasitavat, W., Hoonsopon, D.: Blockchain characteristics and consensus in modern business processes. Journal of Industrial Information Integration **13**, 32–39 (2019)
9. Vasin, P.: Blackcoin's proof-of-stake protocol v2. https://blackcoin.org/blackcoin-pos-protocol-v2-whitepaper.pdf **71** (2014)
10. Larimer, D.: Delegated proof-of-stake (DPoS). Bitshare whitepaper **81**, 85 (2014)
11. Castro, M., Liskov, B., *et al.*: Practical byzantine fault tolerance. In: OsDI, vol. 99, pp. 173–186 (1999)
12. Szabo, N.: Formalizing and securing relationships on public networks. First monday (1997)
13. Lei, A., Cruickshank, H., Cao, Y., Asuquo, P., Ogah, C.P.A., Sun, Z.: Blockchain-based dynamic key management for heterogeneous intelligent transportation systems. IEEE Internet of Things Journal **4**(6), 1832–1843 (2017)
14. Kandi, M.A., Kouicem, D.E., Doudou, M., Lakhlef, H., Bouabdallah, A., Challal, Y.: A decentralized blockchain-based key management protocol for heterogeneous and dynamic IoT devices. Computer Communications **191**, 11–25 (2022)
15. Hameedi, S.S., Bayat, O.: Improving IoT data security and integrity using lightweight blockchain dynamic table. Applied Sciences **12**(18), 9377 (2022)
16. Liu, Q., Luo, L., Wang, J., Li, W., Liu, R., Yu, M.: Key management scheme of distributed IoT devices based on blockchains. IET Communications (2023)
17. Lee, J., Kim, M., Park, K., Noh, S., Bisht, A., Das, A.K., Park, Y.: Blockchain-based data access control and key agreement system in IoT environment. Sensors **23**(11), 5173 (2023)
18. Nafi, M., Bouzefrane, S., Omar, M.: Matrix-based key management scheme for IoT networks. Ad Hoc Networks **97**, 102003 (2020)
19. Nafi, M., Bouzefrane, S., Omar, M.: Efficient and lightweight polynomial-based key management scheme for dynamic networks. In: Mobile, Secure, and Programmable Networking: 6th International Conference, MSPN 2020, Paris, France, October 28–29, 2020, Revised Selected Papers 6, pp. 110–122 (2021). Springer
20. Nafi, M., Messai, M.-L., Bouzefrane, S., Omar, M.: IFKMS: Inverse function-based key management scheme for IoT networks. Journal of Information Security and Applications **71**, 103370 (2022)
21. Mesmoudi, S., Benadda, B., Mesmoudi, A.: SKWN: Smart and dynamic key management scheme for wireless sensor networks. International Journal of Communication Systems **32**(7), 3930 (2019)
22. Saleh, F.: Blockchain without waste: Proof-of-stake. The Review of financial studies **34**(3), 1156–1190 (2021)
23. Johnson, D., Menezes, A., Vanstone, S.: The elliptic curve digital signature algorithm (ECDSA). International journal of information security **1**, 36–63 (2001)
24. Mayer, H.: ECDSA security in bitcoin and ethereum: a research survey. CoinFaabrik, June **28**(126), 50 (2016)
25. Wander, A.S., Gura, N., Eberle, H., Gupta, V., Shantz, S.C.: Energy analysis of public-key cryptography for wireless sensor networks. In: Third IEEE International Conference on Pervasive Computing and Communications, pp. 324–328 (2005). IEEE

Security of Intelligent Transport Systems Using Blockchain: A Review of the Literature

Lamia Hamza, Aida Yachi, and Inès Tigrine

1 Introduction

Blockchain is revolutionizing a number of strategic sectors, including the Internet of Vehicles (IoV), communications and networks, transport applications, payment systems, and energy. It offers innovative solutions for securing data exchanges and transactions in the IoV, ensures transparent and efficient management of information in data management, optimizes financial transactions, and reduces transaction costs. It also facilitates the decentralized and secure management of energy networks and improves traceability and food safety in supply chains. These applications position blockchain as a key technology for transforming the way data and transactions are managed on a global scale. Nowadays, the use of the Internet of Things (IoT) and Vehicular Ad hoc Networks (VANET) is becoming a necessity rather than a choice. We notice this shift in new-generation vehicles equipped with sensors and smart web interfaces like the Web of Things (WoT) which uses open, standardized web technologies at the application layer, such as HTTP, RESTful APIs, JSON, and JavaScript. This approach ensures greater interoperability, making it easier to integrate connected objects with web services and applications, while simplifying their operation and management. This chapter explores key Blockchain technologies applied to the massive data platform for intelligent urban transportation. These technologies aim to solve several major challenges in this field. The unified

L. Hamza (✉)
Laboratory of Medical Informatics and Intelligent and Dynamic Environments, Faculty of Exact Sciences, University of Bejaia, Bejaia, Algeria
e-mail: lamia.hamza@univ-bejaia.dz

A. Yachi · I. Tigrine
Faculty of Exact Sciences, University of Bejaia, Bejaia, Algeria
e-mail: aida.yachi@se.univ-bejaia.dz; ines.tigrine@se.univ-bejaia.dz

S. El Jaouhari, N. El Madhoun (eds.), *Blockchain-based Web of Things Security: Advances, Challenges, and Solutions*, Signals and Communication Technology,
https://doi.org/10.1007/978-3-032-07671-7_5

technology of each node makes it possible to harmonize data from a variety of sources, facilitating the creation of data models tailored to intelligent urban transport. The decentralization offered by Blockchain improves data management by automating supervision and operations, making the system more efficient and reliable than traditional platforms. Compatibility with the Internet of Vehicles (IoV) and vehicle-road collaboration optimizes the use of system resources and enhances road safety. Trusted time stamping is needed to prove that data exists and is correct. Using a registry database makes sure that data cannot be changed, can be tracked, is clear, and can be audited. This makes the Blockchain-based intelligent urban transport system reliable and trustworthy. Despite all these advantages, the application of Blockchain in VANETS and WoT presents a number of challenges, which we will explore through this chapter.

The remainder of this chapter is organized as follows: Sect. 2 presents the background of this study. Section 3 reviews related works. Finally, Sect. 4 concludes the paper and provides some perspectives for future research.

2 Preliminaries

In this section, three concepts related to the review of the literature are outlined: Intelligent Transportation Systems (ITS), Blockchain, and Web of Things (WoT).

2.1 Intelligent Transportation Systems

Daily travel, often hampered by problems such as congestion and accidents, has a negative impact on the economy. ITS uses technology to improve traffic management and optimize travel, thereby transforming transportation networks to provide a better quality of life.

From rudimentary carriages to today's autonomous and connected transport, the evolution of means of transport has been spectacular, responding to technological progress and the needs of users, as shown below:

- **Early studies (1960s–1970s):** The tempting projects of that time were: the research program (CACS) which appeared in Japan. It was the first public-private partnership in the world to have tested an urban area, the ERGS project in the USA, and a similar project (ALI) in Germany. One of the first steps toward intelligent transport management was to focus on traffic management and road safety monitoring, with the introduction of the first Traffic Management Systems (TMS) integrating data on weather, vehicle speed, congestion, and accidents. The data collected is distributed to passengers and the media. These early ITS projects were never implemented due to technical difficulties.

- **Early applications (80s–90s):** After this initial phase, numerous projects were set up. In the 1980s, Europe, realizing that it was lagging behind the USA and Japan in this field, set up a number of programs to develop intelligent transport, such as the PROMETHEUS project (1986) and the DRIVE I programs (1988) and DRIVE II (1991). The USA, for its part, has developed the national IVHS America program. ITS has taken on a global dimension, and the sector's economic opportunities are multiplying and benefiting from funding via public and private research programs. Japan, for its part, has made great efforts to establish a genuine market for ITS and make them commercially available. In this context, it has set up a number of projects, such as the ALV technology in the USA, and the RACS projects in 1984, and the AMTICS project in 1987. All in all, the above-mentioned projects, the result of numerous public and private efforts, have helped to accelerate research in the field of ITS.
- **Major projects (90s–today):** This period was characterized by a worldwide awareness of the need to foster the development of ITS. It was at this time that the first ITS World Congress was held in Paris in 1994. In the same year, and in the USA, the IVHS was upgraded to ITS which reflects the notion of a systemic loop of interaction around four components: the vehicle, the infrastructure, the management system, and the driver. Technological innovations will focus on improving and standardizing transport management systems, integrated driver navigation systems (GPS, traffic management, dashboard), and electronic toll collection systems. Since the 2000s, ITS has become more widespread and modernized with the emergence of smartphones. Mobile applications make it possible to communicate and provide real-time information to passengers and to improve driver assistance and traffic management systems.

Figure 1 represents a schematic that synthesizes the main projects that capture the evolution of ITS over time.

2.2 *ITS Components*

Intelligent transport systems components include electronic communication, data processing, and analysis technologies that enhance the mobility and safety of transport, vehicles, and communication networks:

- **Sensors and detection devices:** Significant advances in sensor technologies have transformed intelligent transportation systems by providing large quantities of reliable data. These versatile sensors make it possible to measure various aspects of vehicles (such as length, weight), analyze traffic (by assessing flow, occupancy, speed), and detect specific events (such as incidents, queues, or violations such as running red lights). This information opens up promising possibilities for precise interventions to improve road safety.
- **Communication systems:** The dissemination of complete data and the collection of information consist in making these data accessible to a variety of users, from

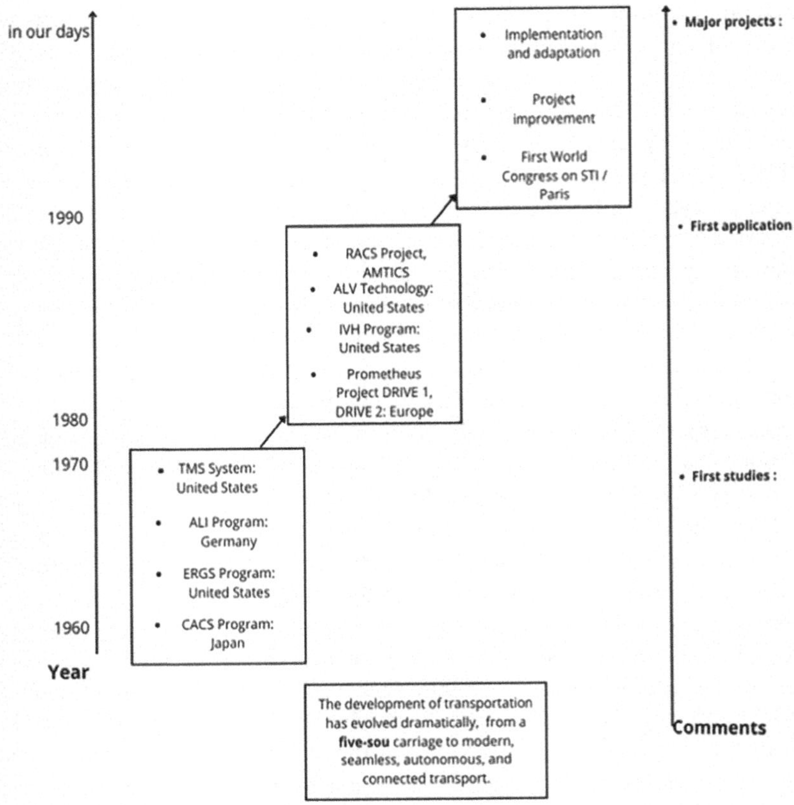

Fig. 1 Evolving historical projects of ITS

the general public (users, passengers, WoT surfers) to professionals (drivers, operating staff, etc.).

- **Localization systems:** Precise tracking of passengers, goods, and vehicles is fundamental to developing more efficient, environmentally friendly, and intelligent mobility. Rapid advances in geolocation services and technologies aim to facilitate access to geographical positions at any time and in any place, thus contributing to better travel management.
- **Cooperative systems:** Collaboration between vehicles and infrastructure goes beyond advances in autonomous systems. In this cooperation, vehicles and infrastructure share their intelligence. Vehicles connect to each other and with road infrastructures to improve the quality of data on their position and the road environment. This information improves existing services and paves the way for new services for road users.

2.3 Blockchain Technology

Blockchain technology initially designed for cryptocurrencies such as Bitcoin, and Blockchain is now essential in a variety of sectors, including ITS. It offers transparency, security, and immutability, improving traffic management, data traceability, and transparency of operations, helping to prevent fraud and optimize the management of transport fleets and services.

The Blockchain can be compared to a vast public register integrating all the exchanges made by its users since its creation. From a more specialized and technical angle, Blockchain refers to a chain of blocks on which information of all kinds is stored. It is generally defined as an information storage and transmission technology that is transparent, secure, and operates without a central control body. It constitutes a database containing the history of all exchanges between its users.

The principles on which Blockchain is based are as follows:

1. **Asymmetric cryptography:** Blockchain uses advanced cryptographic techniques to secure transactions and guarantee data integrity. Asymmetric cryptography, also known as public-key cryptography, is a data encryption and decryption technique that offers a very high level of security, information protection, authenticity, and data confidentiality. In the field of Blockchain, asymmetric cryptography enables a user to sign a transaction on the Blockchain's public register, thereby certifying that he or she is the author. Based on a key pair [1], each user has his or her own key pair (public/private key). The private key remains private to the user, while the public key is accessible to other participants.
2. **Decentralization:** The Blockchain operates without a central authority, and there is no trusted third party, which means that it is distributed across many computers nodes through a network [2].
3. **Immutability:** Data stored on the Blockchain is immutable, which means that once it has been stored, it cannot be modified or deleted without the consensus of the entire network.
4. **Transparence:** All transactions carried out on the Blockchain are generally visible (but encrypted) to all network participants, offering increased transparency.
5. **Traceability:** Although participants operate under pseudonyms, their transactions remain traceable. The complete transaction history can be consulted at any time by all network members. As a decentralized, distributed ledger, Blockchain ensures transparency and immutability, enabling every stage of a specific transaction or activity to be traced.

Blockchain has evolved significantly since its inception, falling into four main stages: Blockchain 1.0, Blockchain 2.0, Blockchain 3.0, and Blockchain 4.0 [3]. Blockchain 1.0, it is worth pointing out that the origins of Blockchain date back to the early 1990s. In 1991, Stuart Haber and W. Scott Stornetta sketched out the architecture of Blockchain, a technology designed to securely time-stamp digital documents to protect them against tampering or back-dating. Their system used a cryptographic Blockchain to store these time-stamped documents. In 1992, the

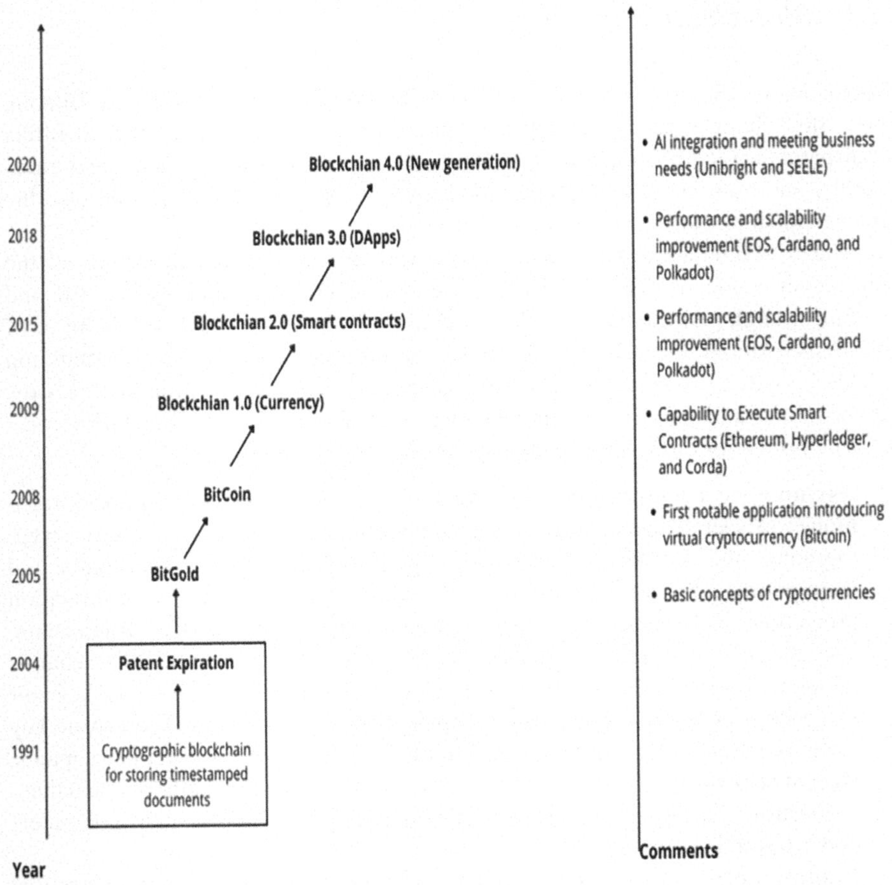

Fig. 2 Diagram illustrating the evolution of blockchain over time

introduction of the Merkle tree improved efficiency by enabling multiple documents to be merged into a single block. However, this technology was neglected, and the patent expired in 2004. In 2005, Nick Szabo developed the fundamental principles of cryptocurrencies and initiated the BitGold project, considered the precursor of Bitcoin [4]. It was only in 2008 that the first practical implementation of a Blockchain became a reality, thanks to the invention of Satoshi Nakamoto, giving rise to Bitcoin, a decentralized peer-to-peer monetary system. Figure 2 represents the evolution of Blockchain technology.

Blockchain is a distributed ledger technology that uses a specific architecture to record and secure transactions. The main elements of its architecture are:

1. **P2P communication protocol:** The P2P model enables each entity to act as both client and server, unlike the client/server model. In this decentralized model, nodes are autonomous and can enter and leave the network freely, while

sharing resources equally [5]. In Blockchain, each node maintains the register, reinforcing security and reliability. Nodes are interconnected, forming a global network that ensures efficient data distribution and enhanced system resilience.

2. **Node:** A node in a P2P network is a machine with the same rights and benefits as the others, with a copy of the distributed registry. Each node validates transactions and blocks, with no master-slave hierarchy.
3. **Transactions:** Transactions, the fundamental elements of a Blockchain, include a recipient address, a sender address, and a value. They have inputs referring to the outputs of previous transactions, thus guaranteeing the traceability of funds. Each transaction modifies the state of the Blockchain and must respect the "ACID" properties (Atomicity, Consistency, Isolation, Durability) to maintain the integrity of the system [5].
4. **Address and wallet:** Addresses in the Blockchain are alphanumeric strings that identify the parties involved in transactions. Users must protect their private keys, which are essential for transactions, and store them in wallets. These wallets, in addition to securing private keys, can also manage public addresses and keys, and calculate the digital assets held by a user.
5. **Blocks:** Blocks are data structures used to group together sets of transactions and distribute them to all network nodes. Each block consists of a header and data. The header contains metadata enabling the validity of the block to be verified, while the data contains the transactions selected by the miner when creating the block. The maximum number of transactions a block can contain depends on the size of the block and the size of each transaction.

   The block header contains several important elements. Firstly, it contains a time stamp indicating the exact date and time when the block was created. Blocks are arranged chronologically, meaning that they follow one another in temporal order. Another key element of the header is the fingerprint (hash) of the previous block. This enables blocks to be linked sequentially. By knowing the previous block's fingerprint, it is possible to verify the integrity of this block and detect any modifications that may have been made. This contributes to the security and immutability of the Blockchain. The block header also contains the root of the Merkle tree. The schematic representation of a block is shown in Fig. 3a. Block information is separated into two distinct parts: the header, which contains information characterizing the block, and the data, which contains information relating to the transactions.
6. **Merkle tree:** The Merkle tree, or hash tree, secures Blockchain data by slicing it into separately hashed elements, then combining these hashes to obtain a unique fingerprint called the Merkle root. The Merkle proof verifies the validity of the data by comparing it to this root and to an authentication path, thus ensuring data integrity.
7. **Chain of Block:** This is the register replicated throughout the network, where all validated blocks are added in an orderly and irreversible fashion, as shown in Fig. 3b. All network members have the same copy of the Blockchain, ensuring the transparency and security of recorded information. If a node tries to introduce an invalid block into the network, it is not validated by most nodes

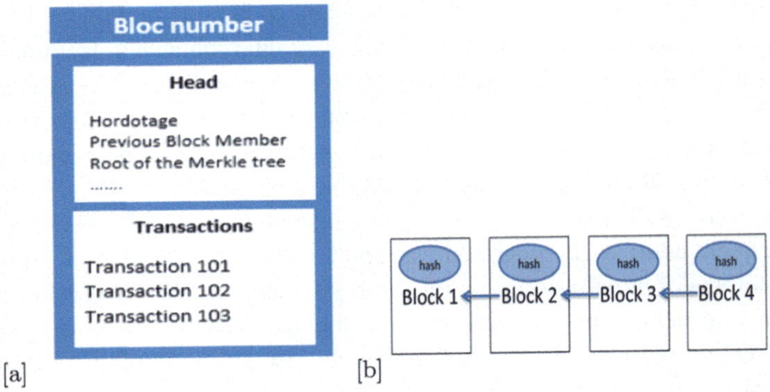

Fig. 3 Example of block and chain of blocks

and is not added to their registry or passed on to their peers. However, it is important to note that in rare cases, some nodes may be corrupted by malicious actors such as hackers or criminal organizations, and could potentially validate this invalid block anyway.

8. **Minors:** Miners verify transactions block by block and are remunerated for their computing power. They collaborate to find proof of work, create and validate new blocks, disseminate transactions, propagate blocks, and confirm transactions, playing a crucial role in the functioning of the blockchain:

   - Adding blocks (mining): Miners compete to solve complex mathematical problems in order to create a new block and obtain a reward. When a miner solves the problem, the block is broadcast on the network for validation by other miners. If consensus is reached, the block is added to the Blockchain and new transactions are recorded.

   - Transaction validation: Miners must validate transactions using consensus algorithms to ensure that all transactions are valid, comply with Blockchain rules, and have not been altered. Once a transaction has been validated, it is added to a block, and this block is added to the Blockchain.

   - Blockchain integrity verification: Blockchain integrity verification is the process by which network nodes ensure that the chain is consistent, with no missing blocks or invalid transactions, by checking each block and the associated proof.

9. **Consensus:** Consensus protocols, essential to Blockchain, guarantee the validity and integrity of transactions in a distributed network using cryptographic mechanisms. Nodes validate blocks, often using methods such as proof of work or proof of stake, before they are added to the Blockchain.

- **Proof of Work (PoW):** It is a key consensus mechanism in the Blockchain and is widely used. Every change in the chain requires subsequent adjustments to maintain validity. The mathematical puzzles forming the proofs of work are complex, security-enhancing challenges. Miners who solve these puzzles first are authorized to publish the block. The probability of success in solving the puzzle is proportional to the miner's contribution. In short, PoW ensures the security and robustness of the Blockchain by aligning incentives with the efforts made by miners.

- **Proof of Stake (PoS):** It is a way of making decisions in a Blockchain based on the possession of cryptocurrency. Unlike Proof of Work (PoW), which is computationally intensive, PoS is not. People who already have tokens can "mine" new blocks in proportion to the quantity of tokens they possess. Let us take a simple example: If someone has 10% of all the tokens in play, that person has around a 10% chance of creating the next block. Miners are chosen according to their wealth rather than their computational capacity, and the one who should create the block does so. If this miner fails to do so in time, another similar miner is chosen to ensure that the process continues. Consensus is based on the Blockchain with the greatest number of tokens in play. PoS draws on game theory to work [6].

10. **Write information request:** This is the first stage of the Blockchain, where agent A asks agent B to write a piece of information. This request is made via a specific software interface connected to the network. Once a write request has been added to a block, it is placed in a queue awaiting consensus validation.
11. **Smart contracts:** The smart contracts are autonomous computer programs that run on a Blockchain when certain predefined conditions are met. They enable transactions and processes to be automated securely and transparently, without the need for intermediaries.

2.4 *Web of Things Architecture*

As we have already pointed out, the concept of the Internet of Vehicles primarily relies on the use of standard web technologies, as opposed to an approach based on specific wireless communications such as vehicle-to-everything (V2X) communication. In order to offer greater flexibility to vehicles, WoT allows easier integration of external services. This allows more effective customization and adaptation to the needs of drivers and vehicle manufacturers. According to W3C [7], the WoT architecture is defined around several fundamental blocks, including the Thing Description (TD), the binding templates, the WoT script API, and the security and confidentiality directives.

- **Description of WoT objects**:
  Thing Description (TD) is a standardized model that describes the operation and capabilities of a connected object using a structured format based on a semantic vocabulary and a JSON representation. TDs provide rich metadata that can be read by humans and interpreted by machines, facilitating interoperability between connected objects. Aligned with the linked data concept, TDs enable information to be organized and used in a structured way. They can be used in the form of raw JSON or enriched with JSON-LD for better interconnection and advanced data processing. A TD describes an object instance using general metadata (name, identifier, description) and may include links to other objects or documents. It also specifies the interaction mechanisms, security requirements, and communication protocols supported. In this sense, it plays a similar role to an `index.html`, serving as a single point of entry for discovering and exploiting the services offered by a connected object.
- **WoT Link Models**:
  The Internet of Things uses a variety of protocols to connect devices, making interoperability a challenge. The Web of Things (WoT) facilitates this compatibility thanks to link models (*binding templates*), which define how objects interact via different protocols (HTTP, MQTT, Modbus), data formats (JSON, CBOR, EXI), and IoT platforms (OPC UA, oneM2M). These models enable an application to extract and use this metadata to establish communication with a connected object. They come in three types:

  - **Protocol links:** network interaction management
  - **Data format links:** structuring and interpreting messages
  - **Platform links:** adaptation to the specific features of IoT platforms

  Binding templates are not strict standards, but serve as guides to ensure compatibility between different IoT systems.
  The linkage models cover three aspects:

  1. **Protocol:** defines the transport protocol used for communication (HTTP, CoAPS, WebSocket) via the access point URI.
  2. **Subprotocol:** Some protocols require specific extensions (e.g., long polling for HTTP) which must be explicitly declared.
  3. **Data format:** Messages can be exchanged in different formats (JSON, CBOR, XML), identified by a MIME type and optional parameters (e.g., `text/plain; charset=utf-8`).

- **WoT Script API**:
  The WoT Script API facilitates the development of IoT applications by enabling JavaScript scripts to be written to control connected objects. Rather than integrating the logic of devices directly into their firmware, which complicates updates, this API enables their behavior to be managed dynamically via reusable scripts. It improves flexibility and productivity by enabling:

  - Move complex tasks to local gateways.

- Bring urgent treatments closer to connected objects instead of sending them to the cloud.

Thanks to this API, scripts can discover, interact with, and publish IoT objects, simplifying integration between different systems and reducing development costs.

– **WoT Security and Confidentiality Guidelines**:
Security is a fundamental aspect of the Web of Things (WoT) and must be taken into account at every stage of the design process. The WoT architecture incorporates several mechanisms to guarantee security and confidentiality, including:

- Security metadata in Thing Descriptions (TDs)
- Separation of responsibilities in the WoT Script API to limit risks
- Specific recommendations for each component of the architecture

3 Related Work

This section reviews previous research and studies on the applications of Blockchain technology in ITS-related fields. We will explore its integration in the Internet of Vehicles (IoV), communications and networks, data management, transportation applications, payment systems, and energy. Figure 4 shows a flow chart of related work.

3.1 IoV

A lot of studies have been carried out in the context of IoV. Yang et al. [8] have introduced a decentralized Blockchain-based platform, underlining its reliability and credibility. This innovative approach is based on a Bayesian inference model, enabling cars to check each other's messages using an intelligent model. Malik

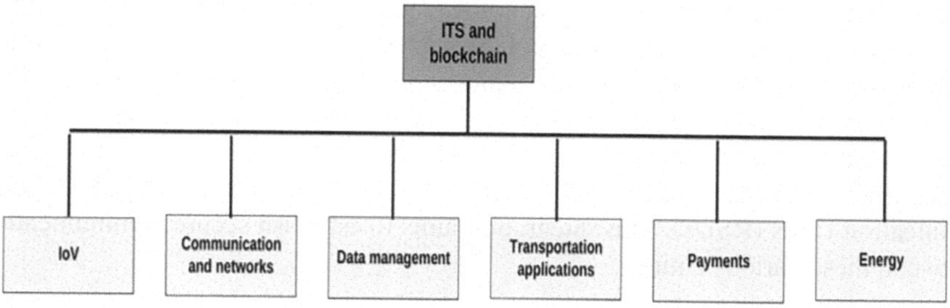

Fig. 4 Application domains of ITS and Blockchain

et al. [9] have developed an intelligent authentication framework for vehicles, leveraging Blockchain functionalities. This system provides updates on vehicle status, while minimizing processing and communication costs compared with a centralized model. In a similar context, Labrador and Hou [10] have proposed a mechanism for authenticating vehicles and data using Blockchain technology. This mechanism enables vehicle and data identification to be authenticated through data packets transmitted from one vehicle to another using Blockchain technology. Rathee et al. [11] have proposed a secure and transparent framework for connected autonomous vehicles (CAVs) using Blockchain to extract and store data. Noh et al. [12] have developed a method for making messages in VANETs more secure and authentic, based on Blockchain with consensus mechanisms. A distributed cloud architecture scheme using Blockchain has been studied by Nadeemet et al. [13] to guarantee the confidentiality of vehicle data and protect it from attack. It is obvious that IoV must include WoT, and more especially, WoV, which is not usually utilized in the literature, such as IoV, because WoV is an extension of WoT applied to vehicles. For this reason, there has been work on WoT, such as [14, 15], but the WoV in relation to Blockchain is still being investigated as an IoV layer. Recently, Hasan et al. [16], Pittaras et al. [17], and Jabbar et al. [18] employed Blockchain in the Internet of Things, but without focusing on WoV as a standalone technology.

3.2 Communication and Networks

In the context of communication and networks, Singh and Kim [19] have advanced an intelligent and reliable communication protocol using Blockchain technology to connect a vehicle system to the cloud. In the near future, Jabbar et al. [20] have developed a decentralized framework based on Blockchain technology which represents a real-time application specification that ensures secure communication between all transport system participants. Wang et al. [21] presented an improved version of the Diffie-Hellman algorithm, designed to strengthen confidence in the verification mechanism applied in a network based on a consortium Blockchain. This enhancement contributes to increasing the reliability of the verification process within the network. At the same time, Nkenyereye et al. [22] presented a hybrid vehicle network, combining 5G technology and the cloud, designed to support an emerging communication protocol dedicated to warning messages. Another significant contribution was made by Singh et al. [23], who have proposed a hybrid Blockchain developed to secure communications between vehicles and between vehicles and infrastructure. Dai et al. [24] have led to the use of Blockchain architecture to connect sensor-rich vehicles, ordinary vehicles, and RoadSide communication Units (RSUs). This integration aims to establish secure communication between these various entities.

3.3 Data Management

There has been a lot of work on data management. Dai et al. [24], Cebe et al. [25], Zhang et al. [26], Sharma et al. [27], Rahman et al. [28], Zhang et al. [29], Shi et al. [30], and Fan et al. [31] focus on the use of smart frameworks and consortium or private Blockchains for the efficient management and sharing of data within various network components, such as vehicles, users, infrastructure nodes, and sensors. Kumar et al. [32] have introduced an access control system to manage traffic data in a network of vehicles using Blockchain technology.

3.4 Transport Applications

There are a variety of transport applications. Hu et al. [33], Jabbar et al. [18, 34], presented Blockchain-based platforms that are finding diverse applications, from smart parking services for cars to vehicle rentals, by Kwame et al. [35], through to the training of autonomous cars by Ghandi and Salvi [36], and up to the establishment of trusted multiparty insurance by Li et al. [37]. These solutions are also used to ensure secure transactions when buying and selling used cars, as suggested by Zhang et al. [38], and to enable transparent dissemination of engine usage history for commercial purposes, as put forward by Masoud et al. [39]. Zhu et al. [40] have developed a smart contract platform for emerging transportation services. This innovative proposition manages driver and vehicle privilege data, storing, sharing, and deleting this information after service completion.

3.5 Payments

Given that the first use of Blockchain was for payment, this leads to work in the context of VANETs. Jeong et al. [41] have envisioned a billing service based on Blockchain technology, ensuring the security of transactions between electric vehicles and charging points. At the same time, Gao et al. [42] have introduced a transaction structure in a Hyperledger system designed to guarantee reliable, tamper-proof payments. On the other hand, Hassidja et al. [43] have contributed to a cost-optimized car rental platform based on intelligent contracts. For a similar purpose, Erdin et al. [44] have designed an innovative topology to personalize Bitcoin transactions in a private network of electric vehicles and charging stations, with the aim of reducing costs and verification times.

3.6 Energy

Research into energy in the IoV is looking at aspects such as smart charging or refueling services within vehicle networks, as examined by Gorenflo et al. [45], Sun et al. [46], Pedrosa et Pau [47], Fu et al. [48], Islam et al. [49], Javed et al. [50], Xu et al. [51], Pajic et al. [52], Wang et al. [53]. These studies explore the integration of Blockchain technology, whether decentralized, private, or consortium-based, to increase the efficiency and security of these services. A concrete example of this application is presented by Pustivsek et al. [54], who have developed a simple charging unit selection mechanism for electric vehicle drivers using smart contracts. In the same context, Zhou et al. [55] were interested in a cost-optimized energy exchange platform running on a consortium Blockchain. This platform innovates by using a contract-based incentive mechanism to encourage more efficient participation of electric vehicles in energy exchange, while respecting their preferences. In parallel, the study conducted by Wang et al. [56] focused on the use of a consensus system based on reputation proof. The aim was to guarantee the security of energy delivery within a network of energy vehicles operating on a private Blockchain.

Table 1 presents a classification of related work in the areas mentioned above and the types of Blockchain applied.

3.7 Discussion

The analysis of the Web of Things underscores the significance of security, particularly in light of the wide variety of linked devices and the increasing dangers posed by cyberattacks of all kinds. The WoT architecture is founded on a number of measures, including the separation of responsibilities in the WoT scripting API and the incorporation of security metadata into Thing Descriptions (TDs). On the other hand, these methods have several shortcomings when it comes to authentication, confidentiality, and access management. Taking this into consideration, Blockchain looks to be a solution that contributes to the enhancement of the security of WoT. It enables rigorous access control, safeguards identity management, and guarantees the integrity of transactions between connected objects because of its decentralized and immutable architecture. Additional benefits include ensuring that identity management is secure. Through the integration of these two technologies, it is possible to create an Internet of Things ecosystem that is more robust, providing transparency and reliability in the process of information transmission. The IoV guarantees V2V and V2X services based on our findings. WoV, which is the web version of IoV, can provide a higher level for external services by ensuring that online services are utilized more effectively. However, this tendency may give rise to new security challenges, which only Blockchain technology can overcome.

Table 1 Classification of related work by areas and Blockchain types

Study	Communication and networks	Data management	Payments	Energy	Type of blockchain
Wang et al. [21]	✓				Consortium Blockchain
Singh et al. [23]	✓				Hybrid Blockchain
Dai et al. [24], Cebe et al. [25], Zhang et al. [26], Sharma et al. [27], Rahman et al. [28], Zhang et al. [29], Shi et al. [30], et Fan et al. [31]		✓			Consortium Blockchain, private Blockchain
Erdin et al. [44]			✓		Decentralized Blockchain
Gorenflo et al. [45], Sun et al. [46], Pedrosa et Pau [47], Fu et al. [48], Islam et al. [49], Javed et al. [50], Xu et al. [51], Pajic et al. [52], Wang et al. [53]				✓	Decentralized Blockchain, consortium Blockchain, private Blockchain
Zhou et al. [55]				✓	Consortium Blockchain
Wang et al. [56]				✓	Private Blockchain

Therefore, the use of Blockchain for WoV holds significant potential as a future research direction.

4 Conclusion

ITS facilitates urban travel by providing transport information and promoting the use of various environmentally friendly modes of travel. Through applications and navigation systems, they simplify trip planning by offering several transport options. They also improve the organization of journeys by connecting different modes of transport. They make travel more accessible for all, especially people with special needs, and improve access to public transport on a social level. Economically, they optimize travel, saving time and money for users while stimulating innovation and job creation in information and communications technologies. With this success, WoV may be more promising than IoV because it offers more benefits for external uses than V2V and V2X. However, existing work on the application of Blockchain for securing VANETs treats WoV as a layer of IoV. Even if the subject of WoV is not yet fully developed, there is a large amount of promise for it to be a future research direction. As for future prospects, we plan to focus on WoV to address interoperability and security issues by integrating Blockchain.

References

1. Jiawen Kang, Zehui Xiong, Dusit Niyato, Ping Wang, Dongdong Ye, Dong In Kim, *Incentivizing consensus propagation in proof-of-stake based consortium blockchain networks, IEEE Wireless Communications Letters*, 8(1):157–160, (2018).
2. Laurent Leloup, *Blockchain: la révolution de la confiance*, Préface de William Mougayar, (2017).
3. Valéria Faure-Muntain, Claude de Ganay, Ronan le Gleut, *Office parlementaire d'évaluation des choix scientifiques et technologiques*, (le 20 juin 2018).
4. Lamia Hamza, Cherifa Azou, & Ilhem Bazizene, Blockchain Technology for Absence Tracking. In: International Conference on Computing and Information Technology. Cham: Springer Nature Switzerland, 2023. p. 10–19.
5. Rateb Jabbar, Eya Dhib, Ahmed Ben Said, Moez Krinchen, Noora Fatais, Esmat Zaidan, Kamel Berkaoui, *Blockchain technology for intelligent transportation systems: A systematic literature review, IEEE Access*, 10:20995–21031, (2022).
6. Matthieu Quiniou, Christophe Debonneuil, *Glossaire blockchain*, Chaire UNESCO ITEN, (avril 2019).
7. World Wide Web Consortium (W3C), Web of Things (WoT) Architecture 1.1, 2020. https://www.w3.org/TR/wot-architecture11
8. Zhe Yang, Kan Yang, Lei Lei, Kan Zheng, Victor CM Leung, *Blockchain-based decentralized trust management in vehicular networks*, IEEE Internet of Things Journal, vol. 6, no. 2, pp. 1495–1505, 2018.
9. Nisha Malik, Priyadarsi Nanda, Arushi Arora, Xiangjian He, Deepak Puthal, *Blockchain based secured identity authentication and expeditious revocation framework for vehicular networks*, in Proceedings of the 2018 17th IEEE International Conference On Trust, Security And Privacy In Computing And Communications/12th IEEE International Conference On Big Data Science And Engineering (TrustCom/BigDataSE), 2018, pp. 674–679.
10. Mirador Labrador, Weiyan Hou, *Implementing blockchain technology in the Internet of Vehicle (IoV)*, in Proceedings of the 2019 International Conference on Intelligent Computing and its Emerging Applications (ICEA), 2019, pp. 5–10.
11. Geetanjali Rathee, Ashutosh Sharma, Razi Iqbal, Moayad Aloqaily, Naveen Jaglan, Rajiv Kumar, *A blockchain framework for securing connected and autonomous vehicles*, Sensors, vol. 19, no. 14, pp. 3165, 2019.
12. Jaewon Noh, Sangil Jeon, Sunghyun Cho, *Distributed blockchain-based message authentication scheme for connected vehicles*, Electronics, vol. 9, no. 1, pp. 74, 2020.
13. Sara Nadeem, Muhammad Rizwan, Fahad Ahmad, Jaweria Manzoor, *Securing cognitive radio vehicular ad hoc network with fog node based distributed blockchain cloud architecture*, International Journal of Advanced Computer Science and Applications, vol. 10, no. 1, pp. 288–295, 2019.
14. Nafaâ JABEUR et Hedi HADDAD, *From intelligent web of things to social web of things*. Facta Universitatis, Series: Electronics and Energetics, 2015, vol. 29, no 3, pp. 367–381.
15. Awais Ahmad, Anand Paul, M. Mazhar Rathhore et al. *Smart cyber society: Integration of capillary devices with high usability based on Cyber–Physical System*. Future Generation Computer Systems, 2016, vol. 56, pp. 493–503.

16. Md Rahat Hasan, Ammar ALAZAB, Siddhartha BarmanJOY, et al. *Smart contract-based access control framework for internet of things devices.* Computers, 2023, vol. 12, no 11, pp. 240.
17. Iakovos Pittaras, Fotiou FOTIOU, Christos Karapapas et al. Secure smart contract-based digital twins for the Internet of Things. Blockchain: Research and Applications, 2024, vol. 5, no 1, pp. 100168.
18. Rateb Jabbar, Noora Fetais, Mohamed Kharbeche, Moez Krichen, Kamel Barkaoui, Mohammed Shinoy, *Blockchain for the Internet of vehicles: How to use blockchain to secure vehicle-to-everything (V2X) communication and payment?* IEEE Sensors Journal, vol. 21, no. 14, pp. 15807–15823, 2021.
19. Madhusudan Singh, Shiho Kim, *Trust bit: Reward-based intelligent vehicle communication using blockchain*, in Proceedings of the IEEE 4th World Forum on Internet of Things (WF-IoT), February 2018, pp. 62–67.
20. Rateb Jabbar, Mohamed Kharbeche, Khalifa Al-Khalifa, Moez Krichen, Kamel Barkaoui, *Blockchain for the Internet of Vehicles: A decentralized IoT solution for vehicle communication using Ethereum*, Sensors, vol. 20, no. 14, pp. 3928, July 2020.
21. Qifan Wang, Lei Zhou, Zhe Tang, Guojun Wang, *A consortium blockchain-based model for data sharing in Internet of Vehicles*, in Proceedings of the International Conference on Smart City Informatization, Springer, 2019, pp. 253–267.
22. Lewis Nkenyereye, Bayu Adhi Tama, Muhammad K. Shahzad, Yoon-Ho Choi, *Secure and blockchain-based emergency driven message protocol for 5G enabled vehicular edge computing*, Sensors, vol. 20, no. 1, pp. 154, December 2019.
23. Priya Singh, Pooja Khanna, Sachin Kumar, *Communication architecture for vehicular ad hoc networks, with blockchain security*, in Proceedings of the International Conference on Computer, Automation, Knowledge Management (ICCAKM), January 2020, pp. 68–72.
24. Yueyue Dai, Du Xu, Ke Zhang, Sabita Maharjan, Yan Zhang, *Deep reinforcement learning and permissioned blockchain for content caching in vehicular edge computing and networks*, IEEE Transactions on Vehicular Technology, vol. 69, no. 4, pp. 4312–4324, 2020.
25. Mumin Cebe, Enes Erdin, Kemal Akkaya, Hidayet Aksu, Selcuk Uluagac, *Block4forensic: An integrated lightweight blockchain framework for forensics applications of connected vehicles*, IEEE Communications Magazine, vol. 56, no. 10, pp. 50–57, 2018.
26. Xiaohong Zhang, Xiaofeng Chen, *Data security sharing and storage based on a consortium blockchain in a vehicular ad-hoc network*, IEEE Access, vol. 7, pp. 58241–58254, 2019.
27. Sachin Sharma, Kamal Kumar Ghanshala, Seshadri Mohan, *Blockchain-based internet of vehicles (IoV): An efficient secure ad hoc vehicular networking architecture*, in Proceedings of the 2019 IEEE 2nd 5G World Forum (5GWF), 2019, pp. 452–457.
28. Abdur Rahman, Mamunur Rashid, Stuart J. Barnes, Sayed Maruf Abdullah, *A blockchain-based secure internet of vehicles management framework*, in Proceedings of the 2019 UK/China Emerging Technologies (UCET), 2019, pp. 1–4.
29. Lei Zhang, Mingxing Luo, Jiangtao Li, Man Ho Au, Kim Kwang Raymond Choo, Tong Chen, Shengwei Tian, *Blockchain based secure data sharing system for Internet of Vehicles: A position paper*, Vehicular Communications, vol. 16, pp. 85–93, April 2019, https://doi.org/10.1016/j.vehcom.2019.01.004.
30. Kexin Shi, Liehuang Zhu, Can Zhang, Lei Xu, Feng Gao, *Blockchain-based multimedia sharing in vehicular social networks with privacy protection*, Multimedia Tools and Applications, vol. 79, no. 11, pp. 1–21, 2020, https://doi.org/10.1007/s11042-019-08603-y.
31. Kai Fan, Qiang Pan, Kuan Zhang, Yuhan Bai, Shili Sun, Hui Li, Yintang Yang, *A secure and verifiable data sharing scheme based on blockchain in vehicular social networks*, IEEE Transactions on Vehicular Technology, vol. 69, no. 6, pp. 5826–5835, June 2020, https://doi.org/10.1109/TVT.2020.2981028.
32. Amritesh Kumar, Amrendra Singh Yadav, Dharmender Singh Kushwaha, *VChain: Efficient blockchain-based vehicular communication protocol*, in Proceedings of the 10th International Conference on Cloud Computing, Data Science and Engineering (Confluence), 2020, pp. 762–768.

33. Quanyi Hu, Simon Fong, Peng Qin, Jingzhi Guo, Yong Zhang, Dan Xu, Yuanlan Chen, Jerome Yen, *Intelligent car parking system based on blockchain processing reengineering*, in Advances in E-Business Engineering for Ubiquitous Computing: Proceedings of the 16th International Conference on e-Business Engineering (ICEBE 2019), 2020, pp. 265–273, Springer.
34. Rateb Jabbar, Moez Krichen, Mohammed Shinoy, Mohamed Kharbeche, Noora Fetais, Kamel Barkaoui, *A model-based and resource-aware testing framework for parking system payment using blockchain*, in Proceedings of the 2020 International Wireless Communications and Mobile Computing (IWCMC), 2020, pp. 1252–1259.
35. Opuni-Boachie Obour Kwame, Qi Xia, Emmanuel Boateng Sifah, Sandro Amofa, Kingsley Nketia Acheampong, Jianbin Gao, Ruidong Chen, Hu Xia, James C Gee, Xiaojiang Du, et al., *V-chain: A blockchain-based car lease platform*, in Proceedings of the 2018 IEEE International Conference on Internet of Things (iThings) and IEEE Green Computing and Communications (GreenCom) and IEEE Cyber, Physical and Social Computing (CPSCom) and IEEE Smart Data (SmartData), 2018, pp. 1317–1325.
36. Gaurav Meera Gandhi, *Artificial intelligence integrated blockchain for training autonomous cars*, in Proceedings of the 2019 Fifth International Conference on Science Technology Engineering and Mathematics (ICONSTEM), vol. 1, pp. 157–161, 2019.
37. Zengxiang Li, Zhe Xiao, Quanqing Xu, Ekanut Sotthiwat, Rick Siow Mong Goh, Xueping Liang, *Blockchain and IoT data analytics for fine-grained transportation insurance*, in Proceedings of the 2018 IEEE 24th International Conference on Parallel and Distributed Systems (ICPADS), 2018, pp. 1022–1027.
38. Jinyu Zhang, Honghui Zhao, Yumeng Yang, Jiaqi Yan, *Towards transparency and trustworthy: A used-car deposit platform based on blockchain*, in Proceedings of the 2019 IEEE 19th International Conference on Software Quality, Reliability and Security Companion (QRS-C), 2019, pp. 46–50.
39. Mohammad Masoud, Yousef Jaradat, Ismael Jannoud, Dema Zaidan, *CarChain: A novel public blockchain-based used motor vehicle history reporting system, 2019 IEEE Jordan International Joint Conference on Electrical Engineering and Information Technology (JEEIT)*, 683–688, (2019), IEEE.
40. Yishui Zhu, Feng Du, Bo Wu, Zongtao Duan, *A sharing platform of emergency cars based on blockchain environment, Frontier Computing: Theory, Technologies and Applications (FC 2019)*, 8, 199–209, (2020), Springer.
41. Seohyeon Jeong, Nhu-Ngoc Dao, Yunseong Lee, Cheol Lee, Sungrae Cho, *Blockchain-based billing system for electric vehicle and charging station*, Proceedings of the 10th International Conference on Ubiquitous Future Networks (ICUFN), July, 308–310, (2018).
42. Feng Gao, Lienhuang Zhu, Meng Shen, Kashif Sharif, Zhiguo Wan, Kui Ren, *A blockchain-based privacy-preserving payment mechanism for vehicle-to-grid networks*, IEEE Network, 32(6), 184–192, Nov./Dec., (2018), DOI: 10.1109/MNET.2018.1800104.
43. Vikas Hassija, Mohd Zaid, Gurjot Singh, Amit Srivastava, Vikas Saxena, *Cryptober: A blockchain-based secure and cost-optimal car rental platform*, Proceedings of the 12th International Conference on Contemporary Computing (IC3), August, 1–6, (2019).
44. Enes Erdin, Mumin Cebe, Kemal Akkaya, Senay Solak, Eyuphan Bulut, Selcuk Uluagac, *Building a private Bitcoin-based payment network among electric vehicles and charging stations, Proceedings of the IEEE International Conference on Internet of Things (iThings), IEEE Green Computing and Communications (GreenCom), IEEE Cyber-Physical-Social Computing (CPSCom), IEEE Smart Data (SmartData)*, July, 1609–1615, (2018).
45. Christian Gorenflo, Lukasz Golab, Srinivasan Keshav, *Mitigating trust issues in electric vehicle charging using a blockchain*, in Proceedings of the Tenth ACM International Conference on Future Energy Systems, pp. 160–164, 2019.
46. Haoli Sun, Song Hua, Ence Zhou, Bingfeng Pi, Jun Sun, Kazuhiro Yamashita, *Using ethereum blockchain in Internet of Things: A solution for electric vehicle battery refueling*, in Blockchain–ICBC 2018: First International Conference, Held as Part of the Services Conference Federation, SCF 2018, Seattle, WA, USA, June 25–30, 2018, pp. 3–17, Springer.

47. Alejandro Ranchal Pedrosa, Giovanni Pau, *ChargeltUp: On blockchain-based technologies for autonomous vehicles*, in Proceedings of the 1st Workshop on Cryptocurrencies and Blockchains for Distributed Systems, pp. 87–92, 2018.
48. Zhengtang Fu, Peiwu Dong, Yanbing Ju, *An intelligent electric vehicle charging system for new energy companies based on consortium blockchain*, Journal of Cleaner Production, vol. 261, p. 121219, 2020, Elsevier.
49. Mohammad Mainul Islam, Mohammad Shahjalal, Mohammad Khalid Hasan, Yeong Min Jang, *Blockchain-based energy transaction model for electric vehicles in v2g network*, in Proceedings of the 2020 International Conference on Artificial Intelligence in Information and Communication (ICAIIC), pp. 628–630, 2020, IEEE.
50. Muhammad Umar Javed, Nadeem Javaid, Abdulaziz Aldegheishem, Nabil Alrajeh, Muhammad Tahir, Muhammad Ramzan, *Scheduling charging of electric vehicles in a secured manner by emphasizing cost minimization using blockchain technology and IPFS*, Sustainability, vol. 12, no. 12, p. 5151, 2020, MDPI.
51. Zisang Xu, Wei Liang, Kuan-Ching Li, Jianbo Xu, Hai Jin, *A blockchain-based roadside unit-assisted authentication and key agreement protocol for internet of vehicles*, Journal of Parallel and Distributed Computing, vol. 149, pp. 29–39, 2021, Elsevier.
52. Jelena Pajic, José Rivera, Kaiwen Zhang, Hans-Arno Jacobsen, *EVA: Fair and auditable electric vehicle charging service using blockchain*, Proceedings of the 12th ACM International Conference on Distributed and Event-based Systems, 262–265, (2018).
53. Zhishang Wang, Mark Ogbodo, Huakun Huang, Chen Qiu, Masayuki Hisada, Abderazek Ben Abdallah, *AEBIS: AI-enabled blockchain-based electric vehicle integration system for power management in smart grid platform*, IEEE Access, 8, 226409–226421, (2020), IEEE.
54. Matevž Pustišek, Andrej Kos, Urban Sedlar, *Blockchain based autonomous selection of electric vehicle charging station, 2016 International Conference on Identification, Information and Knowledge in the Internet of Things (IIKI)*, 217–222, (2016), IEEE.
55. Zhenyu Zhou, Bingchen Wang, Yufei Guo, Yan Zhang, *Blockchain and computational intelligence inspired incentive-compatible demand response in internet of electric vehicles*, IEEE Transactions on Emerging Topics in Computational Intelligence, 3(3), 205–216, (2019), IEEE.
56. Yuntao Wang, Zhou Su, Ning Zhang, *BSIS: Blockchain-based secure incentive scheme for energy delivery in vehicular energy network*, IEEE Transactions on Industrial Informatics, 15(6), 3620–3631, (2019), IEEE.

A Blockchain Solution for Reliable and Secure Distributed SSO Applications

Oliver Bell and Hisham Al-Assam

1 Introduction

Managing user identities and access credentials has become increasingly complex in the era of interconnected devices and digital services. This challenge is especially pronounced in Web of Things (WoT) networks, where seamless interactions among devices, users, and applications are crucial for ensuring optimal functionality. However, traditional identity management systems often fail to meet the needs these dynamic environments. Their inherent limitations compromise security while hindering usability, scalability, and network efficiency, posing significant obstacles to the advancement of WoT ecosystems.

One of the primary challenges users face today is "password fatigue"—the mental burden of managing numerous unique passwords for different accounts and services. In the context of Web of Things (WoT) networks, where interconnected platforms and services are vast, this issue is further magnified. A unified authentication mechanism could mitigate insecure practices such as password reuse, the creation of weak passwords, or unsafe credential storage [1]. These behaviors significantly heighten the risk of credential theft and unauthorized access, threatening individual accounts and potentially compromising entire networks.

O. Bell · H. Al-Assam (✉)
The University of Buckingham, Buckingham, UK
e-mail: hisham.al-assam@buckingham.a.cuk

S. El Jaouhari, N. El Madhoun (eds.), *Blockchain-based Web of Things Security: Advances, Challenges, and Solutions*, Signals and Communication Technology,
https://doi.org/10.1007/978-3-032-07671-7_6

Traditional centralized authentication solutions exacerbate the problem, requiring users to repeatedly authenticate across various devices and services. This fragmented experience undermines the seamless interactions that WoT networks are designed to provide. Moreover, administrators face the daunting task of managing diverse credentials, enforcing access policies, and maintaining security compliance across multiple systems, leading to inefficiencies, higher operational costs, and increased security vulnerabilities.

Without Single Sign-On (SSO), users must authenticate individually with each service, device, or application within a network. This lack of integration results in a fragmented user experience, decreased productivity, and an increased risk of errors in credential management. Moreover, the absence of a centralized mechanism for enforcing security policies makes the network more vulnerable to cyberattacks.

A poorly configured or centralized SSO system introduces additional risks [2], including:

1. **Single points of failure**: Centralized SSO systems rely on a single authentication server or provider. If this server is compromised or experiences an outage, it can lead to widespread disruption, locking users out of all connected applications or exposing sensitive data.
2. **Scalability challenges**: As networks grow, traditional SSO systems often struggle to handle the increased load, resulting in latency, bottlenecks, and reduced reliability.
3. **Privacy risks**: Centralized SSO providers, such as social logins from Google or Facebook, raise concerns over data misuse and surveillance, as they have access to extensive user activity data across platforms.
4. **Administrative burdens**: Traditional SSO systems require continuous maintenance, updates, and monitoring to comply with evolving security standards, adding to the complexity and cost of managing WoT networks.

These limitations are particularly problematic in Web of Things (WoT) ecosystems, where security, efficiency, and seamless interoperability are essential. In such environments, devices often operate autonomously, sharing sensitive data and performing tasks without direct human intervention. A compromised identity or a poorly implemented authentication mechanism within these networks could lead to severe consequences, such as widespread security breaches, data leaks, or even physical harm, depending on the nature of the connected devices and their functions.

The Promise of Blockchain-Based Decentralized SSO

A blockchain-based decentralized Single Sign-On (SSO) system presents a transformative solution to the challenges of traditional identity management. Unlike conventional systems, blockchain decentralizes authentication by distributing credentials and identity data across a network of nodes. This approach removes the reliance on a single server or authority, significantly reducing the risk of single points of failure and enhancing resilience against cyberattacks. Key benefits of a decentralized SSO system include [3]:

1. **Enhanced security**: By leveraging the blockchain's immutable, tamper-proof ledger, decentralized SSO systems ensure that credentials cannot be altered or deleted without detection. The use of cryptographic algorithms further safeguards user data from unauthorized access.
2. **User autonomy**: Decentralized systems empower users to control their credentials and authentication data. This user-centric approach not only promotes privacy but also aligns with privacy regulations like the General Data Protection Regulation (GDPR), fostering trust in the authentication process.
3. **Scalability and efficiency**: By distributing the authentication workload across nodes, decentralized networks can scale efficiently as the network grows. This is particularly beneficial in Web of Things (WoT) environments, where billions of interconnected devices need seamless authentication.
4. **Seamless interoperability**: A blockchain-based SSO system can unify authentication across diverse devices and services, ensuring a smooth user experience while maintaining stringent security protocols.
5. **Regulatory compliance and transparency**: Blockchain's transparent ledger provides an auditable trail of authentication events, ensuring accountability and facilitating compliance with industry standards and regulations.

Adopting a decentralized SSO system can have profound impacts on WoT networks, where secure and efficient identity management is critical. It mitigates password fatigue, enhances user convenience, and reduces the administrative burdens associated with traditional systems. Furthermore, it aligns with the decentralized and distributed nature of WoT, creating a more secure and user-friendly ecosystem. Beyond WoT, such systems hold the potential to revolutionize identity management across industries, ushering in an era of seamless and secure authentication.

By addressing the limitations of existing systems, a blockchain-based decentralized SSO framework lays a solid foundation for securing the rapidly evolving WoT landscape and beyond. This chapter delves into the implementation of such a system, illustrating its ability to meet the demands of modern networks and its potential for broader adoption.

This chapter is structured as follows: Section 1.1 provides background on the vulnerabilities of centralized identity management systems and introduces blockchain technology as a transformative solution. Section 2 presents the proposed blockchain-based SSO solution, detailing its architecture, key components, and authentication workflow. Section 3 discusses the proof-of-concept implementation and evaluates the system's performance and security. Section 4 compares the proposed solution with traditional SSO systems and explores its potential applications in various industries. Finally, Section 5 outlines future directions and concludes the chapter by summarizing the key contributions and implications of blockchain-based SSO systems for WoT networks and beyond.

2 Background

The increasing complexity and interconnectedness of digital ecosystems, including Web of Things (WoT) networks, have exposed the limitations of traditional identity management systems. These systems often struggle to address critical concerns such as security, scalability, and user control. While initially popularized by cryptocurrencies like Bitcoin and Ethereum, blockchain technology has evolved into a transformative solution for secure authentication and decentralized identity management. Its application now extends beyond financial transactions to tackle challenges in a variety of domains, including WoT networks.

2.1 Vulnerability of Centralized Identity Management

Traditional centralized systems are vulnerable to becoming single points of failure, leaving sensitive user data exposed to breaches. As shown in Fig. 1, major data breaches have compromised billions of records, eroding trust in centralized solutions. Blockchain addresses these risks by distributing data across a decentralized network of nodes, reducing reliance on any single entity and ensuring tamper-proof authentication records. This decentralization not only fosters trust but also enhances resilience, minimizing the impact of breaches.

The rise in cybercrime underscores the urgent need for more robust identity management systems. A recent survey published by the UK government revealed that over half of UK businesses experienced breaches or attacks in 2024 [5].

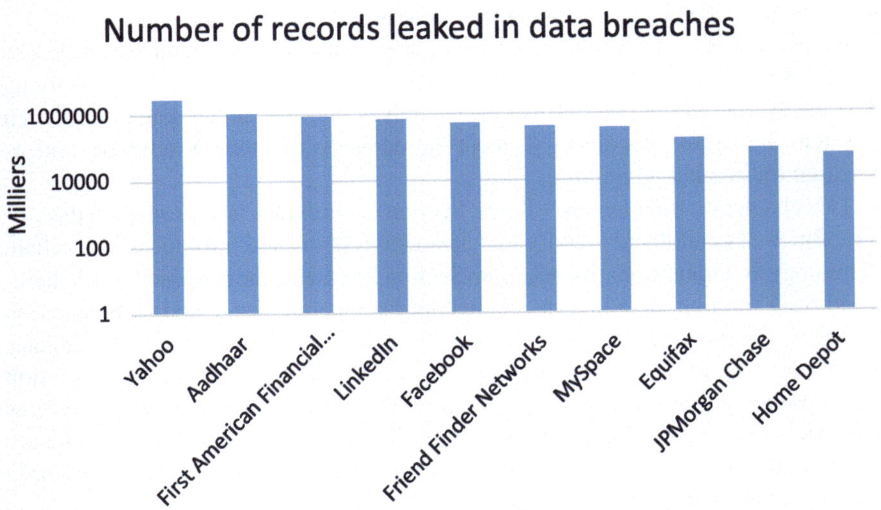

Fig. 1 Ten most significant data breaches—Data from [4]

Centralized identity systems, as prime targets for attackers, often result in large-scale compromises. Blockchain-based systems, on the other hand, allow users to control their data directly, eliminating single points of failure. This decentralized, user-centric approach strengthens security, mitigates data theft risks, and increases resilience against cyberattacks.

2.2 *Transforming Identity Management for WoT Networks*

WoT networks, which consist of billions of interconnected devices, demand secure, scalable, and distributed identity solutions. Blockchain technology aligns with the decentralized nature of WoT by offering:

- **Tamper-proof data**: Immutable records ensure the integrity of credentials and transactions.
- **Scalability**: Decentralized systems handle authentication efficiently across expanding networks.
- **User empowerment**: Users maintain control over their credentials, enhancing privacy and security.

The decentralized nature of blockchain enhances traditional Single Sign-On (SSO) systems and transforms identity management within WoT networks [6]. By providing secure and transparent authentication, blockchain fosters trust among users, administrators, and devices, paving the way for the broader adoption of decentralized technologies.

2.3 *Single Sign-On: Structure, Benefits, and Challenges*

Single Sign-On (SSO) is an authentication mechanism that allows users to access multiple applications with a single set of credentials, simplifying the user experience and alleviating the mental burden of managing multiple passwords—a phenomenon commonly known as password fatigue [7]. SSO eliminates repetitive logins in enterprise and Web of Things (WoT) environments, where users interact with various devices and services, thereby enhancing both usability and security. Traditional SSO systems typically include the following components [8]:

- **User interface**: The entry point where users submit their credentials, usually via a web-based login page.
- **Authentication server**: Validates the credentials against a stored database (e.g., LDAP or Active Directory). Upon successful validation, it issues a session token.
- **Service providers**: These applications trust the authentication server and grant users' access to services without requiring repeated logins, relying on session tokens.

- **Token issuer**: Generates secure tokens that encapsulate authentication data (e.g., identity and permissions), typically with time limitations for enhanced security.

Traditional SSO offers several advantages [9]. It improves the user experience by streamlining authentication processes and reducing login redundancies. It simplifies administrative tasks through centralized access management, making it easier to apply security policies across multiple applications. Additionally, SSO enhances security by reducing the number of passwords users need to manage, which lowers the risk of weak or reused passwords. Furthermore, centralizing access logs simplifies auditing and reporting, aiding in compliance with regulations.

On the other hand, traditional SSO systems face several challenges [10]. A compromised or unavailable authentication server can lead to a single point of failure, disrupting access to all connected services. As the system expands to accommodate more users and applications, performance can degrade, leading to scalability issues. Centralized data storage also raises privacy concerns, as it increases the risk of misuse or surveillance of user data. Furthermore, integrating SSO with existing systems requires significant resources and ongoing maintenance.

2.4 *Introduction to Blockchain Technology*

Blockchain technology offers a decentralized approach to managing data, addressing many of the limitations of traditional SSO systems. By recording transactions on an immutable distributed ledger, blockchain ensures transparency, security, and resilience against vulnerabilities like single points of failure [11].

2.4.1 Blockchain and Identity Management

In the context of identity management, blockchain decentralizes authentication by distributing credentials across a network of nodes. This eliminates the need for a central authority, enhances security through cryptographic techniques, and provides tamper-proof audit trails [12]. These features are particularly suited to WoT networks, where decentralization and scalability are crucial

2.4.2 Challenges of Blockchain

Despite its advantages, blockchain technology faces several challenges. Scalability is an issue, as consensus mechanisms across nodes can slow transaction processing in large-scale systems. Implementing blockchain also requires specialized expertise, computational resources, and robust security protocols, which adds to its complexity and costs. Additionally, storing sensitive data on public blockchains may conflict with privacy regulations, such as the General Data Protection Regulation (GDPR) [13].

2.5 Self-Sovereign Identities (SSI)

Self-Sovereign Identity (SSI) marks a significant shift in digital identity management, emphasizing user ownership and control of credentials. Unlike centralized and federated models, SSI eliminates intermediaries, allowing individuals to manage and share their identity data securely [14]. SSI is based on several key principles [15]:

- User control: Individuals determine when and with whom to share their credentials.
- Decentralization: Identity data is stored across a blockchain network, eliminating the need for central authorities.
- Selective disclosure: Users can share only the necessary information for authentication, thus protecting their privacy.
- Interoperability: SSI systems are designed to work seamlessly across platforms and applications.

In WoT ecosystems, SSI provides secure, decentralized identity management across devices, reducing reliance on central authorities and aligning with the distributed nature of interconnected systems.

3 The Proposed Blockchain-Based Single Sign-On Solution

This section outlines the design and implementation of a blockchain-based SSO system and explores its implications for enhancing security in Web of Things (WoT) networks. The primary objectives of the proposed solution are as follows:

- **Address single points of failure**: Develop a decentralized authentication mechanism that distributes user data across a blockchain network, improving resilience against system-wide failures.
- **Enhance privacy and security**: Leverage blockchain's cryptographic features to secure user data, ensuring tamper-proof authentication and reducing the risks of identity theft and data breaches.
- **Enable user autonomy**: Empower users with control over their credentials, ensuring compliance with modern privacy regulations like GDPR and building trust in digital identity management systems.
- **Align with WoT networks**: Design a solution that is compatible with the decentralized structure of WoT networks, enabling seamless authentication across interconnected devices and platforms.

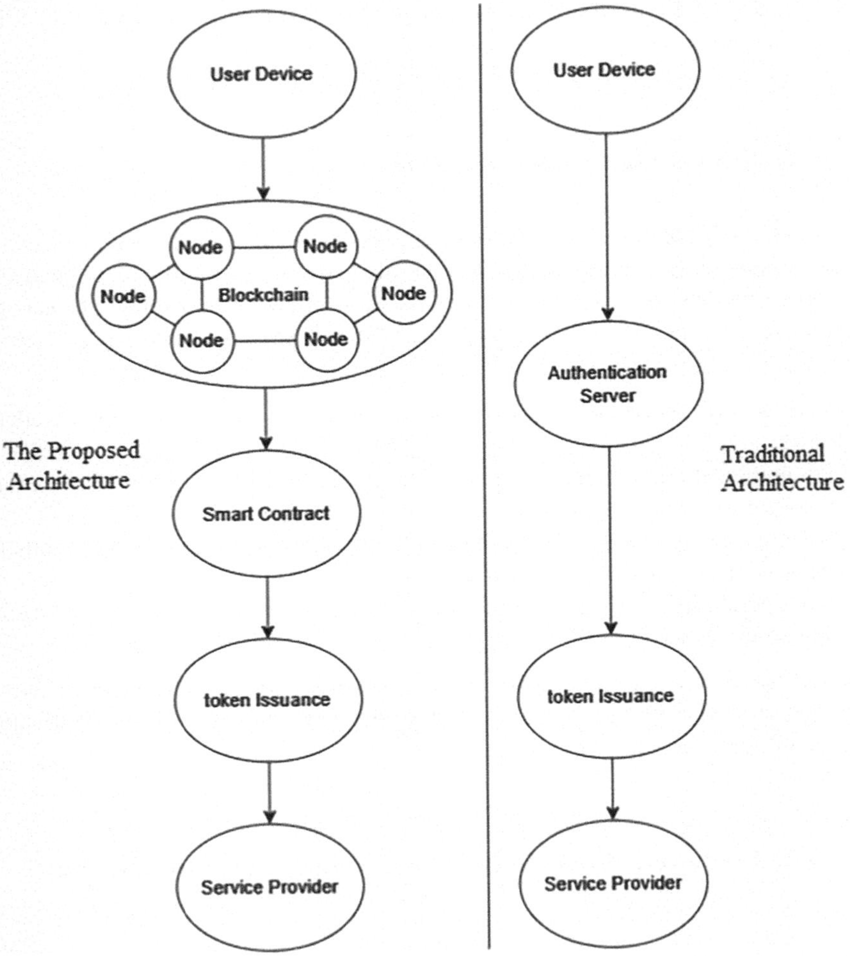

Fig. 2 The proposed architecture vs traditional architecture

- **Provide transparent and scalable solutions**: Utilize blockchain's transparency and scalability to create an SSO framework that meets the growing demands of various industries.

This section details the blockchain-based SSO solution as illustrated in Fig. 2, key components, the authentication workflow, identity management, and critical security considerations. The system is designed to leverage blockchain's decentralized, transparent, and immutable nature to overcome the limitations of traditional SSO systems. The architecture of the proposed solution consists of three interconnected layers:

1. **Front-end interface**: The user-facing application that handles user registration, login, and session management. This layer provides the interface through which users interact with the system, enabling seamless authentication and access to services.
2. **Back-end server**: A Django-based application (using Python as the website back-end framework) that facilitates communication between the front end and the blockchain. This server is responsible for managing user sessions, issuing authentication tokens, and ensuring smooth interaction between the user interface and the blockchain network.
3. **Blockchain network**: A decentralized ledger that stores identity data and authentication logs. This network is powered by smart contracts, which automate the verification and validation of authentication processes while ensuring transparency, immutability, and security of the authentication records.

3.1 The Key Components of the Blockchain SSO Solution

The solution has front end and back end as shown in Fig. 3. The front end is the primary user interaction point for registration, login, and identity management while the back end bridges the user interface and the blockchain.

The system should offer a user-friendly front-end interface for seamless registration and credential management. It leverages secure cryptographic operations to verify user identity and integrates seamlessly with back-end web framework, such as Django, for authentication requests and session management.

The back-end server is responsible for:

- Token management: Issues, validates, and revokes session tokens based on authentication events [16].
- Smart contract interaction: Communicates with the blockchain to verify user credentials and log authentication events [17].
- Session management: Tracks active sessions and enforces expiration or revocation policies for enhanced security.

Additionally, the back end implements:

- Role-Based Access Control (RBAC): Assigns permissions to users based on their roles, ensuring granular access control [18].
- Secure communication**:** Uses HTTPS and cryptographic libraries to protect data transmission between components.

The **blockchain** serves as the foundation of the SSO framework, providing decentralized credential storage, transparent authentication logging, and automated processes via smart contracts. User credentials are securely hashed and stored immutably on the blockchain [19]. Every authentication event is recorded transparently, creating an auditable trail of activities. Smart contracts automate

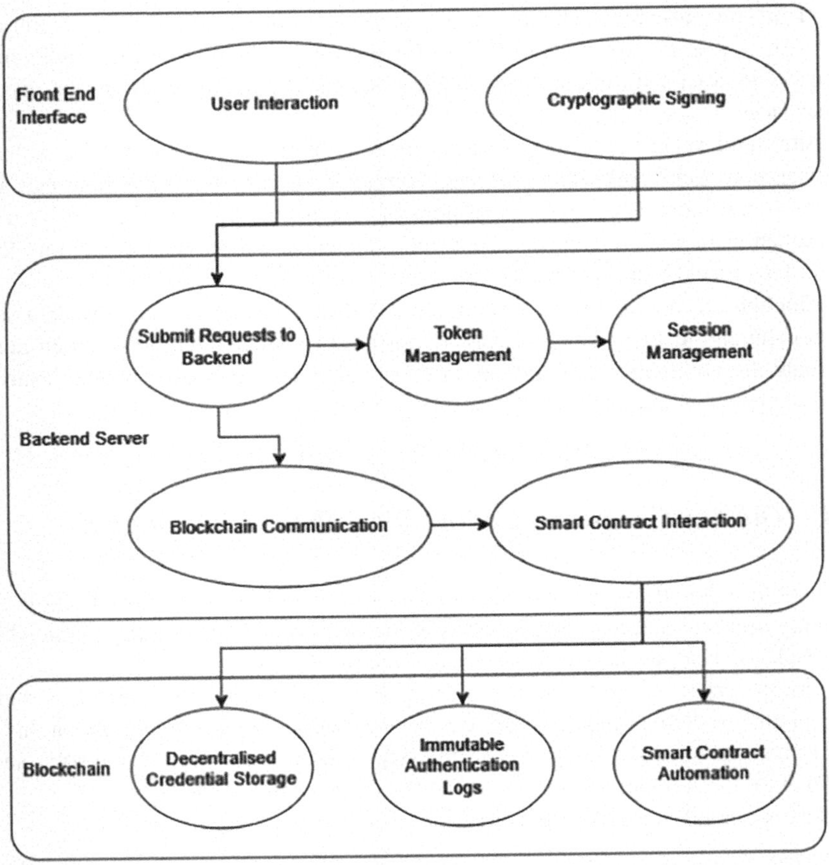

Fig. 3 Key components of the blockchain-based SSO

critical authentication processes, including verifying user signatures, issuing session tokens, and managing token revocation or expiration.

3.2 *Identity Management and Authentication Workflow*

The blockchain-based SSO system prioritizes security, efficiency, and transparency in its identity management and authentication workflow illustrated in Fig. 4.

User Registration: Users initiate registration through the front-end interface by providing minimal information (e.g., email or username). The system then generates a unique blockchain address and a corresponding private/public key pair for the user. To ensure data integrity and immutability, credential information is hashed and securely stored on the blockchain via a smart contract.

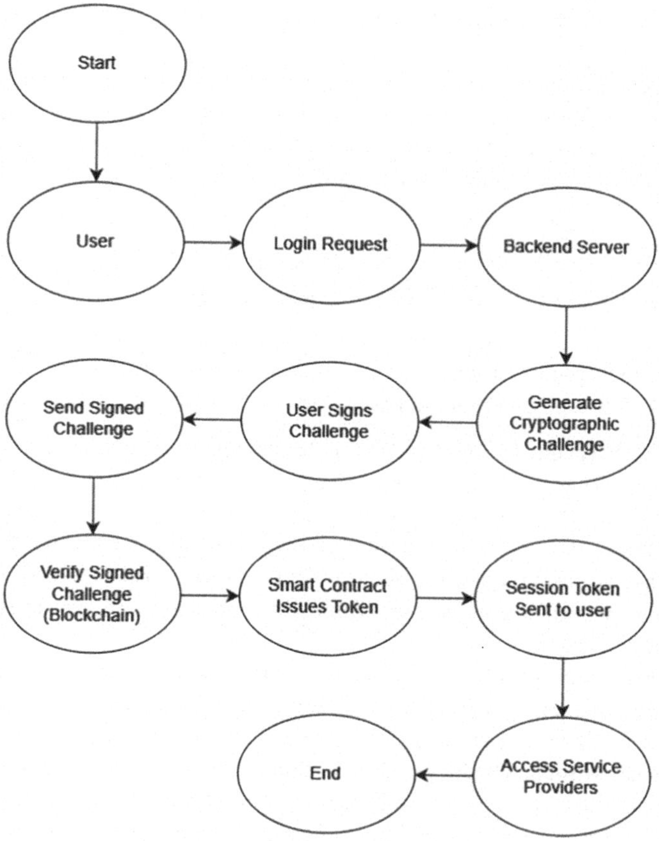

Fig. 4 Authentication workflow in the proposed solution

Key Features of the Workflow:

- **Anonymity:** User identity is represented by blockchain addresses, ensuring privacy.
- **Transparency:** Authentication events are immutably logged, providing verifiable records.
- **Efficiency:** Smart contract automation reduces latency and administrative overhead [20].

3.2.1 Login and Authentication Process

Users initiate the login process by submitting their username or blockchain address. The back-end system generates a cryptographic challenge, which the user signs using their private key. This signed challenge is then transmitted to the blockchain, where a smart contract verifies the signature against the user's public key stored on the chain.

Upon successful verification, the smart contract issues a session token and logs the authentication event. This session token is presented to service providers, who validate it using cryptographic standards, granting users seamless access to authorized applications without requiring repeated logins.

3.3 Security and Scalability Considerations

The system incorporates robust security measures with several components illustrated in Fig. 5 to mitigate potential vulnerabilities via the following four points.

- Role-**Based Access Control (RBAC):** Implements granular permissions based on user roles (e.g., admin, user, guest), ensuring that only authorized individuals can access specific resources and reducing the risk of privilege escalation.
- **Encryption:** All communication between the front end, back end, and blockchain is encrypted using secure protocols (e.g., HTTPS and TLS). User credentials and tokens are cryptographically protected to prevent interception or unauthorized access.
- **Anonymity and privacy:** Blockchain addresses function as pseudonymous identifiers, preserving user privacy. The system employs selective disclosure techniques, allowing users to share only necessary credentials.
- **Tamper resistance:** The blockchain's immutable ledger guarantees that authentication records cannot be altered, providing a reliable history of events. Smart contracts enforce consistent and tamper-proof authentication processes.

Scalability Through Layer 2: As the number of users and transactions on the blockchain grows, scalability becomes a critical concern. To address this, the system integrates Layer 2 solutions (e.g., rollups) to improve transaction throughput and reduce latency in large-scale environments. Layer 2 solutions are protocols that operate on top of the main blockchain, processing transactions off-chain to improve scalability and reduce fees. Rollups are a specific type of Layer 2 solution that bundles multiple transactions into a single transaction on the main chain, significantly increasing throughput [21].

4 Proof-of-Concept Implementation and Results

This section **outlines** the technical implementation of the blockchain-based SSO proof-of-concept and evaluates its performance and security through testing. By detailing the development environment, key components, and results, we aim to showcase the practical feasibility and potential of the proposed solution.

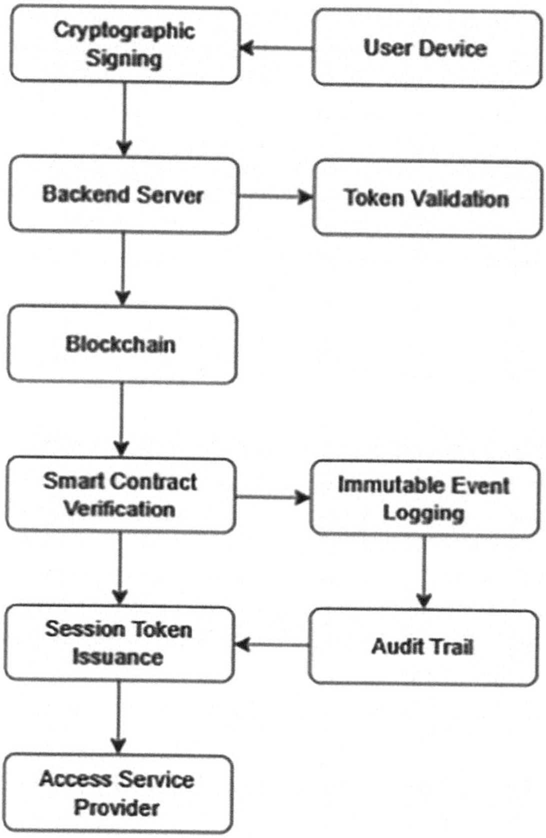

Fig. 5 Data flow for blockchain-based SSO security

4.1 Implementation Details

The blockchain-based SSO system was developed using a robust combination of tools, frameworks, and technologies to ensure functionality, security, and scalability.

- **Front-end interface:** Built with React, HTML, CSS, and JavaScript, it offers a dynamic, user-friendly platform for registration and login. Users interact with the interface to input minimal data (e.g., username or email) and perform cryptographic operations securely, ensuring private keys remain on their devices.
- **Back-end application server:** Developed using Python with Django, the back-end handles session management and communicates with the Ethereum blockchain via the Web3.py library. It also implements role-based access control for dynamic permission allocation.
- **Blockchain platform:** The system leverages the Ethereum Virtual Machine (EVM) with the Truffle suite for smart contract testing.

- **Smart contracts:** Written in Solidity, the contracts manage decentralized credential storage, cryptographic signature verification, and token issuance. Authentication events are logged immutably for transparency and auditability.
- **Testing framework:** Truffle was used to simulate and validate smart contract functionality in a controlled environment.

By integrating these components, the system provides secure and decentralized authentication, aligning with modern privacy regulations and the needs of Web of Things (WoT) networks.

4.2 Development Challenges

Gas costs on Ethereum were a significant challenge, necessitating the optimization of smart contract functions to minimize gas consumption during authentication and logging processes [22]. Latency was another concern, particularly during signature verification, and efforts were made to reduce delays caused by blockchain interactions to ensure a smoother user experience. Key management was a crucial design consideration, focusing on creating secure and user-friendly mechanisms for storing and managing private keys, balancing ease of use with robust security to mitigate risks.

4.3 Performance, Security, and Privacy Evaluation

The blockchain-based SSO system demonstrated strong performance, security, and privacy capabilities. The average login time, including blockchain interactions, was approximately 1.5 s, while token validation latency with service providers was around 300 milliseconds. Stress testing revealed the system could handle up to 1000 concurrent login requests with minimal performance degradation, though large-scale deployments may require Layer 2 scaling solutions to maintain efficiency.

Security validation confirmed the robustness of the system against various attack vectors. Immutable and tamper-resistant blockchain-based credential storage ensured verifiable authentication records. Replay attacks were mitigated by employing unique session tokens for each login, while cryptographic signatures and secure communication channels effectively prevented man-in-the-middle (MITM) attacks. Additionally, reliance on private key-based authentication eliminated the risks of credential stuffing, as static passwords were no longer required.

User privacy was also a key focus, with blockchain addresses providing pseudonymity to safeguard sensitive user data during authentication. These measures collectively highlight the system's capacity to deliver high performance, robust security, and enhanced privacy in real-world applications.

5 Discussions

5.1 Traditional SSO vs. Proposed Blockchain-Based SSO for WoT

When compared to traditional SSO systems, the blockchain-based SSO solution demonstrated significant advantages in security, privacy, and user empowerment. Table 1 compares traditional SSO and the proposed blockchain-based SSO for WoT. While traditional systems remain effective for small-to-medium-scale applications, the blockchain-based system showcased its potential to address modern demands for decentralization and regulatory compliance. Table 1 highlights the blockchain-based SSO's potential to redefine secure, scalable, and privacy-preserving authentication frameworks, especially in environments demanding decentralization, such as the Web of Things (WoT).

In addition to this high-level comparison above, it is essential to assess how various blockchain-based SSO solutions in the literature differ in terms of their contributions, challenges, and adaptability to WoT environments. Table 2 offers a detailed comparative analysis of these solutions, highlighting their key contributions, the challenges they address, proof-of-concept implementations, user autonomy, scalability, and relevance to WoT applications.

Table 1 Traditional SSO vs. proposed blockchain-based SSO for WoT

Feature	Traditional SSO	Blockchain-based SSO
Security	Vulnerable to single points of failure and tampering due to centralized data storage	Highly resilient due to decentralized data storage and cryptographic signature verification
Scalability	Effective for small-to-medium-scale applications, but performance degrades with large-scale implementations	Manages small-to-medium scale efficiently; requires Layer 2 solutions for optimal performance in large-scale
User empowerment	Central authority retains control over user credentials; limited compliance with modern privacy regulations	Grants users control over their credentials, ensuring compliance with data privacy regulations like GDPR
Privacy	Relies on centralized storage of sensitive user data, increasing the risk of misuse or breaches	Pseudonymous blockchain addresses safeguard sensitive data during authentication
Resilience to attacks	Prone to replay attacks, man-in-the-middle attacks, and credential stuffing due to reliance on static passwords and centralized systems	Mitigates attacks through unique session tokens, cryptographic methods, and elimination of static passwords
Transparency	Limited visibility into authentication processes, with no immutable audit trails	Offers transparent, tamper-proof audit trails using blockchain's immutable ledger

Table 2 Comparative analysis of blockchain-based SSO solutions

Ref	Key contributions	Challenges addressed in traditional SSOs	Proof of concept implementation	User's privacy	Scalability	WoT tailoring
[10]	Blockchain as a solution to SSO vulnerabilities, enhancing security	Centralization risks, single point of failure, and trust issues	Theoretical focus, lack of practical details	Not covered	Not covered	No
[23]	Blockchain-based SSO concept for improved trust and decentralization	Trust, transparency, and decentralization	Theoretical focus, lack of practical details	Not covered	Limited discussion	No
[22]	Blockchain-based SSO system for web applications	Security and usability of SSO in web applications	Yes, a prototype was provided	Not covered	Limited discussion	No
[21]	Blockchain-based SSO solution tailored for IoT ecosystems	Scalability and security challenges in IoT environments	Yes but high resource consumption for IoT device	Not covered	Yes	No
[24]	Blockchain-based federated identity system for secure authentication	Centralized control and privacy issues in federated identity systems	Theoretical framework; no implementation	**Yes**	Not covered	No
[25]	Blockchain for decentralized identity federations, enhancing privacy	Centralization and privacy concerns in identity federations	Conceptual only and lacks practical considerations	**Yes**	Not covered	No
[26]	Privacy-preserving authentication method using zero-knowledge proofs	Privacy and security in blockchain authentication	Complex cryptographic techniques with potential performance overhead	**Yes**	Not covered	No
Our proposal	User-centric approach tailored for WoT aligns with modern privacy regulations and emphasizes **user autonomy and privacy**	Privacy, security, and of SSO in blockchain for WoT applications	Yes, detailed architecture with three layers	Yes	Layer 2 solution (rollups) to address scalability issues	**Yes**

As observed in Table 2, existing blockchain-based SSO solutions vary in scope and implementation. While several studies focus on theoretical frameworks, some provide proof-of-concept implementations demonstrating the feasibility of blockchain-based authentication systems. Notably, our proposed solution differentiates itself by incorporating a user-centric approach aligned with modern privacy regulations, such as GDPR, while addressing scalability issues through Layer 2 rollup solutions. Furthermore, our model is explicitly designed for WoT applications, ensuring seamless authentication in decentralized and resource-constrained environments.

By analyzing these comparisons, it becomes evident that while blockchain-based SSO has made significant strides, challenges such as scalability, regulatory compliance, and practical deployment remain key considerations. The proposed approach bridges these gaps by integrating robust privacy-preserving mechanisms with scalability solutions tailored for WoT ecosystems.

5.2 *Other Potential Applications of Blockchain-Based SSO*

Blockchain-based SSO demonstrates versatility in addressing security, efficiency, and regulatory challenges across diverse industries, offering transformative potential in sectors such as healthcare and finance. This section highlights its applications in these domains as case studies.

In healthcare, strong identity management solutions are essential for safeguarding sensitive patient information, ensuring privacy compliance, and enabling seamless.

interactions among patients, providers, and institutions [23]. Blockchain-based SSO allows patients to securely manage and selectively share their medical records with healthcare providers, enhancing privacy, data integrity, and access to critical.

medical history. By enabling secure authentication across hospitals, laboratories, and insurance providers, it fosters interoperability while reducing administrative burdens through automated access permissions via smart contracts. Additionally, the transparent and auditable nature of blockchain authentication logs simplifies compliance with regulations such as HIPAA and GDPR, ensuring patient data is accessed only by authorized entities and maintaining accountability.

In the financial sector, challenges such as fraud prevention, identity theft, and regulatory compliance require innovative identity management solutions [25]. Blockchain-based SSO provides a decentralized framework that enhances security and user control. It enables secure and tamper-proof identity verification for banking and financial services, significantly reducing risks associated with identity theft and fraud. Decentralized authentication ensures the secure execution of financial transactions without exposing user credentials to centralized servers. Additionally, blockchain SSO simplifies compliance with Anti-Money Laundering (AML) and Know Your Customer (KYC) regulations by maintaining an immutable record of authentication events and enabling secure sharing of verified identity attributes.

5.3 *Challenges of Blockchain-Based SSO Adaptation*

Despite its advantages, blockchain-based SSO adoption faces several significant challenges that must be addressed for widespread implementation.

One major concern is scalability. Blockchain networks often experience latency and throughput limitations due to the computational overhead of consensus mechanisms. These issues become especially critical in high-volume environments, such as large-scale IoT networks, where authentication requests can surge. Optimization strategies, including the implementation of Layer 2 solutions like rollups or sharding, are essential to enhance scalability and maintain performance under heavy loads.

Regulatory compliance presents another critical challenge. While blockchain ensures data immutability and transparency, this characteristic can conflict with data privacy regulations such as the General Data Protection Regulation (GDPR). For instance, GDPR mandates the right to erase personal data, which is difficult to achieve on an immutable ledger. Even when user credentials are stored as hashes, concerns persist about meeting legal requirements. Techniques such as zero-knowledge proofs and off-chain storage solutions offer potential pathways to address these regulatory conflicts while preserving blockchain's security and transparency benefits.

Additionally, adoption barriers remain a significant obstacle. Many organizations are unfamiliar with blockchain technology and may view it as complex or costly to implement compared to traditional systems. Concerns about the learning curve, integration challenges, and the need for specialized expertise often deter adoption. Overcoming these barriers requires education, interoperability standards, and user-friendly tools that simplify the transition to blockchain-based systems.

This chapter argues that by tackling these challenges, blockchain-based SSO can unlock its full potential and achieve broader adoption across diverse industries.

6 Future Directions and Conclusion

Blockchain-based Single Sign-On (SSO) systems are poised to revolutionize identity management by offering a secure, transparent, and decentralized alternative to traditional frameworks. This chapter has explored their potential to address critical challenges in various domains, including healthcare, finance, and the Web of Things (WoT). While the advantages of blockchain-based SSO are clear, achieving widespread adoption will require addressing technical, regulatory, and usability hurdles.

6.1 Future Directions

To fully realize the transformative potential of blockchain SSO, several key areas warrant further exploration:

- **Scalability enhancements**: Layer 2 solutions such as rollups, sidechains, and state channels can address the latency and throughput limitations of blockchain networks. These technologies will enable blockchain SSO systems to scale effectively for high-demand environments like WoT networks.
- **Privacy preservation**: Zero-knowledge proofs (ZKPs) present an opportunity to enhance privacy by allowing users to prove their credentials without revealing sensitive information. Incorporating ZKPs will align blockchain SSO systems with stringent data privacy regulations such as GDPR, building greater trust among users.
- **Interoperability and standards**: The heterogeneous nature of WoT networks requires standardized approaches to integration. Developing interoperability protocols for blockchain SSO will ensure seamless operation across diverse devices and platforms, fostering greater adoption.
- **Decentralized key management**: Private key management remains a cornerstone of blockchain security. Innovations such as threshold cryptography and multi-signature schemes can mitigate risks associated with key loss, providing users with secure and user-friendly recovery options.
- **Dynamic access control**: Leveraging blockchain smart contracts for real-time, fine-grained access control policies will enhance flexibility and security. Such capabilities are particularly valuable in dynamic environments like healthcare, where access requirements frequently evolve.
- **Integration with emerging technologies**: Combining blockchain SSO systems with artificial intelligence for adaptive threat detection and leveraging ultra-low latency 5G/6G networks can significantly enhance system resilience and efficiency. These integrations can expand the applicability of blockchain SSO across emerging digital ecosystems.
- **Regulatory alignment**: Engaging with regulators to define compliance frameworks will address governance, data sovereignty, and privacy concerns. Establishing clear guidelines will accelerate trust and adoption in both public and private sectors.
- **User experience improvements**: Simplifying the user interface and creating intuitive tools for managing credentials will be crucial for mainstream acceptance. Developing user-friendly wallets and seamless platforms will lower the barriers to entry for non-technical users.

6.2 Conclusion

This chapter explained the transformative potential of blockchain-based SSO systems in redefining identity management for distributed ecosystems. By decentralizing authentication processes and leveraging blockchain's inherent immutability and transparency, these systems mitigate traditional SSO vulnerabilities, such as single points of failure and privacy risks, while addressing scalability challenges. For WoT networks, blockchain-based SSO provides a decentralized framework to secure interactions across interconnected devices and users.

However, realizing this vision requires overcoming challenges in scalability, regulatory compliance, and user adoption. Future developments in Layer 2 scaling, privacy-preserving technologies, decentralized key management, and interoperability standards will be pivotal. Moreover, integrating complementary technologies and fostering collaboration with regulators will lay the groundwork for a secure, user-centric digital ecosystem.

By addressing these challenges and capitalizing on emerging opportunities, blockchain-based SSO systems can establish themselves as a cornerstone of reliable and secure distributed identity management. As the digital world grows increasingly interconnected, these systems will play a critical role in fostering trust, efficiency, and resilience across diverse applications.

References

1. Al-Slais, Y., & El-Medany, W.M. (2022). User-centric adaptive password policies to combat password fatigue. International Arab Journal of Information Technology, 19(1), 55–62.
2. Roy, S., Matloob, S. & Mukhopadhyay, D. (2021, October). On application of Blockchain to Enhance Single Sign-On (SSO) systems. In *2021 IEEE 20th International conference on trust, security and privacy in computing and communications (TrustCom)* (pp. 1191–1195). IEEE.
3. Ferrag, M.A., Derdour, M., Mukherjee, M., Derhab, A., Maglaras, L., & Janicke, H. (2018). Blockchain technologies for the Internet of Things: Research issues and challenges. IEEE Internet of Things Journal, 6(2), 2188–2204.
4. NordVPN. "*Biggest Data Breaches in History*." Retrieved from https://nordvpn.com/blog/biggest-data-breaches/
5. Department for Science, Innovation, Technology (2024). "*Cyber security breaches survey*", Retrieved from https://www.gov.uk/government/statistics/cyber-security-breaches-survey-2024
6. Sousa, P.R., Resende, J.S., Martins, R. & Antunes, L. (2022). The case for blockchain in IoT identity management. Journal of Enterprise Information Management, 35(6), 1477–1505.
7. Arslan, H. & Aslan, H. (2019, April). Blockchain-based single sign-on support for IoT environments. *2019 27th Signal Processing and Communications Applications Conference (SIU)*, 1–4. IEEE.
8. Parkin, S., Krol, K., Becker, I. & Sasse, M.A. (2016). Applying cognitive control modes to identify security fatigue hotspots. *Twelfth Symposium on Usable Privacy and Security (SOUPS 2016)*.

9. Johnson, A.D., Alom, I., & Xiao, Y. (2023, November). Rethinking single sign-on: A reliable and privacy-preserving alternative with verifiable credentials. *Proceedings of the 10th ACM Workshop on Moving Target Defense,* 25–28.
10. Matloob, S. (2019). Exploring applicability of blockchain to enhance Single Sign-On (SSO) systems. University of North Florida.
11. Baars, D.S. (2016). Towards self-sovereign identity using blockchain technology (Master's thesis). University of Twente.
12. Ferdous, M.S., Chowdhury, F., & Alassafi, M.O. (2019). In search of self-sovereign identity leveraging blockchain technology. IEEE Access, 7, 103059–103079.
13. Da Xu, L., & Viriyasitavat, W. (2019). Application of blockchain in collaborative Internet-of-Things services. IEEE Transactions on Computational Social Systems, 6(6), 1295–1305.
14. Raggett, D. (2015). The Web of Things: Challenges and opportunities. Computer, 48(5), 26–32.
15. Bakre, A., Patil, N., & Gupta, S. (2017). Implementing decentralized digital identity using blockchain. International Journal of Engineering Technology Science and Research, 4(10), 379–385.
16. Aldweesh, A. (2023). The Intersection of Blockchain and The Web of Things: A Systematic Mapping Study. Journal of Namibian Studies: History Politics Culture, 35, 3937–3959.
17. Nokhbeh Zaeem, R., Chang, K.C., Huang, T.C., Liau, D., Song, W., Tyagi, A., Khalil, M., Lamison, M., Pandey, S., & Barber, K.S. (2021, December). Blockchain-based self-sovereign identity: Survey, requirements, use-cases, and comparative study. *IEEE/WIC/ACM International Conference on Web Intelligence and Intelligent Agent Technology*, 128–135.
18. Zhang-Kennedy, L., Chiasson, S., & van Oorschot, P. (2016, June). Revisiting password rules: Facilitating human management of passwords. *2016 APWG Symposium on Electronic Crime Research (eCrime)*, 1–10. IEEE.
19. Durneva, P., Cousins, K., & Chen, M. (2020). The current state of research, challenges, and future research directions of blockchain technology in patient care: Systematic review. Journal of Medical Internet Research, 22(7), e18619.
20. Conti, M., Kumar, E.S., Lal, C., & Ruj, S. (2018). A survey on security and privacy issues of Bitcoin. IEEE Communications Surveys & Tutorials, 20(4), 3416–3452.
21. Arslan, H., & Aslan, H. (2019). Blockchain Based Single Sign-On Support for IoT Environments. *27th Signal Processing and Communications Applications Conference (SIU)*, Sivas, Turkey, pp. 1–4.
22. Tanrıverdi, M. (2020). Design and Implementation of Blockchain Based Single Sign-On Authentication System for Web Applications. Sakarya University Journal of Computer and Information Sciences, 3(3), 343–354.
23. Balatska, V.S., Poberezhnyk, V., Petriv, P., & Opirskyy, I. (2024). Blockchain Application Concept in SSO Technology Context. In Proceedings of the 2024 International Conference on Computer Science, Information Technology, and Systems (CSITS), 38–49.
24. Gilda, S., Jain, T., & Dhalla, A. (2023). None Shall Pass: A Blockchain-Based Federated Identity Management System. Inventive Computation and Information Technologies, Lecture Notes in Networks and Systems, 563, 329–352.
25. Shuhan, M.K.B., Hasnayeen, S.M., Das, T.K., Sakib, M.N., & Ferdous, M.S. (2023). Decentralised Identity Federations Using Blockchain. International Journal of Information Security, 23, 1–20.
26. Chalkias, K., Roy, S., & Gorbunov, S. (2024). zkLogin: Privacy-preserving Blockchain authentication with existing credentials. *Proceedings of the 2024* ACM SIGSAC Conference on Computer and Communications Security (CCS '24), USA

Building Trust in the Web of Things: Decentralized Identity Management with Blockchain

Souhail Mssassi and Anas Abou El Kalam

1 Introduction

The Web of Things (WoT) represents an evolution of the Internet of Things (IoT) that emphasizes web standards to seamlessly interconnect the digital and physical worlds. In the WoT ecosystem, billions of smart devices—from home thermostats and wearable health monitors to industrial sensors and autonomous vehicles—collect and exchange data to enable automated processes, data-driven decisions, and personalized user experiences [1]. This vast scale and heterogeneity of devices make identity management a critical concern [2]. As WoT deployments grow, ensuring each device can be uniquely identified, securely managed, and reliably authenticated is essential for preserving trust and integrity across the network [3]. However, traditional identity management systems in IoT have largely been centralized. Typically, a single authority or a small group of Identity Providers (IdPs) issues and manages credentials for devices, often backed by directories and standardized authentication protocols like OAuth or SAML. While such centralized approaches worked for earlier, smaller-scale IoT networks, they struggle to scale for the rapidly expanding WoT. Relying on a central IdP introduces bottlenecks and single points of failure: if the IdP or its database is compromised or goes offline, the security of the entire WoT ecosystem is at risk. Moreover, performance suffers as more devices connect—a lone IdP can become overwhelmed, leading to higher latency and lower reliability in authentication services.

Centralized systems also raise privacy and control issues: device owners have limited agency over their identities and must entrust sensitive data to third parties, which may misuse aggregated data or face breaches [4]. The lack of interoperability

S. Mssassi (✉) · A. A. El Kalam
University of Cadi Ayyad, Marrakech, Morocco
e-mail: souhail.mssassi@owasp.org; a.abouelkalam@uca.ma

S. El Jaouhari, N. El Madhoun (eds.), *Blockchain-based Web of Things Security: Advances, Challenges, and Solutions*, Signals and Communication Technology,
https://doi.org/10.1007/978-3-032-07671-7_7

between proprietary identity silos and the burden of complying with varying regulations (such as GDPR in Europe or CCPA in California) further highlight the shortcomings of traditional identity frameworks. These challenges underscore the need for a more decentralized solution to identity management in the WoT. By removing reliance on any single authority, a decentralized approach can eliminate the single point of failure, improve scalability through distributed operation, and enhance user privacy and control. Blockchain technology has emerged as a promising foundation for such solutions, offering an immutable, distributed ledger that can securely record device identities and credentials without a central administrator. In a blockchain-based identity system, devices could self-authenticate and self-authorize their interactions via smart contracts on the ledger, reducing or even eliminating the need for traditional centralized IdPs. Techniques from the Self-Sovereign Identity (SSI) paradigm—including Decentralized Identifiers (DIDs) for devices and cryptographic Verifiable Credentials (VCs)—allow device owners to retain control over identity information and share only what is necessary with service providers, enhancing privacy by design.

In this chapter, we present a blockchain-based decentralized identity management framework tailored for the Web of Things. Our goal is to demonstrate how integrating blockchain with SSI principles and DIDs can address the scalability, security, and privacy limitations of centralized WoT identity systems. We introduce a new framework that redistributes identity control to device owners and automates trust decisions using smart contracts, thereby building greater trust in WoT ecosystems. We also discuss how this approach mitigates key vulnerabilities in current WoT deployments and outline the challenges that remain.

The remainder of this chapter is organized as follows. Section 2 provides foundational concepts of identity management, first defining identity, credentials, and verification in the WoT context, then reviewing traditional centralized identity management architectures and their limitations. Section 3 introduces blockchain-based decentralized identity management. It outlines how blockchain's properties address WoT challenges and describes core components of the decentralized approach, namely decentralized identifiers, self-sovereign identity principles, and smart contracts. Section 4 presents the proposed framework for decentralized identity management in WoT: an overview of the framework, its key components and processes, and an illustrative example demonstrating the framework in action. Section 5 discusses the challenges of implementing decentralized identity in WoT at scale and potential solutions. Finally, Section 6 concludes the chapter and highlights directions for future work in building trustful and scalable WoT identity systems.

2 Foundations of Identity Management

2.1 Understanding Identity in the WoT Context

In the context of the Web of Things, identity refers to the distinguishing information that defines a device or entity within the network. At minimum, a device's identity includes a unique identifier (such as a numeric ID, URI, or name) and often a set of attributes or metadata (device type, owner, capabilities, etc.) that describe it. Managing identities involves issuing credentials that assert certain information about the device (for example, a digital certificate or token vouching for a device's model or permissions) and establishing mechanisms to verify those credentials when the device interacts with other entities. In essence, identity, credentials, and verification are three pillars of trust in WoT:

1. **Identity**: A unique representation of a device (or user) in the system, often encompassing attributes that need to be trusted.
2. **Credentials**: Digital attestations or proofs linked to an identity. These could be certificates, cryptographic keys, or tokens that prove a device's identity or privileges. For instance, a manufacturer might issue a credential certifying a sensor's calibration, or a user's controller might hold a token proving it is authorized to command a smart appliance.
3. **Verification**: The process of checking credentials and identities when a device attempts an action. This ensures the device *is who it claims to be* and has the rights it claims to have. Verification may involve cryptographic checks (digital signature validation), querying an identity directory or contacting an identity provider.

In WoT, robust identity management is fundamental to security. Without reliable identities, malicious agents could impersonate devices or manipulate data, undermining the integrity of the system. Each device must therefore be uniquely identified and authenticated before it can transmit data or execute commands in the network. Proper identity management allows the network to confidently attribute actions and data to the correct sources, enabling accountability and fine-grained access control [5]. For example, a smart door lock should only accept open/close commands from a verified homeowner's device; a sensor reading should be trusted only if it comes from an authenticated sensor. By managing identities and credentials, the WoT can enforce such security policies and maintain trust among heterogeneous devices. In short, identity management serves as a *cornerstone* for securing IoT/WoT devices—it underpins authentication (verifying who a device is) and authorization (what the device is allowed to do), thereby helping prevent unauthorized access and ensuring data originates from legitimate sources.

2.2 *Traditional Centralized Identity Management*

Traditional IoT identity management architectures have been largely centralized, mirroring frameworks used in enterprise networks. In a centralized model, a trusted authority (or a small set of authorities) is responsible for handling identity records and authentication decisions for all devices. Figure 1 depicted a layered view of such a system.

The key components typically include:

- **Identity Provider**: A central server or service that creates and maintains identities. The IdP issues credentials (e.g., digital certificates, tokens, or login credentials) to devices or users and vouches for their identity when requested.
- **Directory/repository**: A database or directory where identity information and credentials are stored. This repository contains device attributes, public keys or certificates, and possibly access control policies linked to each identity [6].
- **Authentication/authorization server**: The component that devices contact to prove their identity when they join or act in the network (see Fig. 2). For example, a centralized authentication server might handle login requests or cryptographic challenge-response protocols, and an authorization server or module decides if an authenticated device has permission for a requested action. Common protocols used here include OAuth 2.0, OpenID Connect, and SAML [7], which facilitate token-based authentication and single sign-on in centralized setups. In IoT scenarios, proprietary or simplified versions of these protocols may be used to accommodate constrained devices.
- **Access control mechanisms**: Once identities are verified, access control lists or policies (potentially based on models like Role-Based Access Control, RBAC—

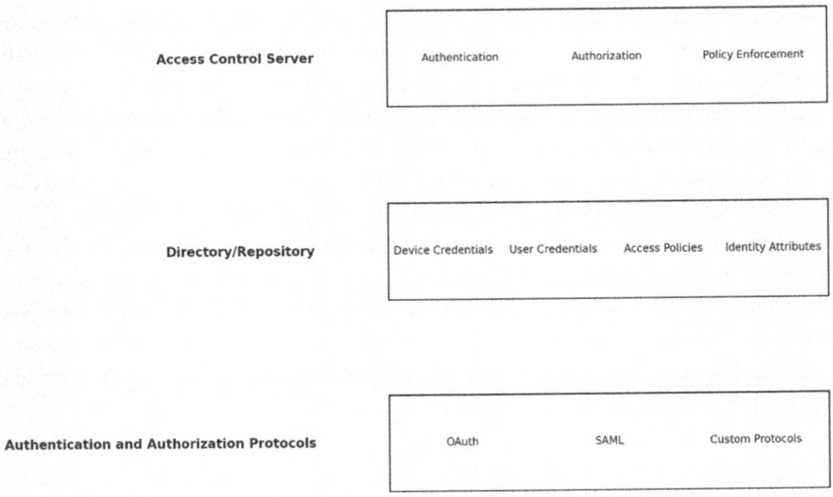

Fig. 1 Layered architecture of a centralized identity management model

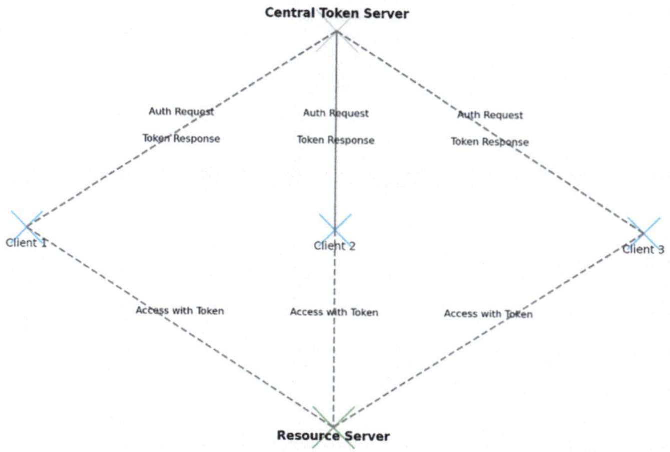

Fig. 2 OAuth-based centralized token server model

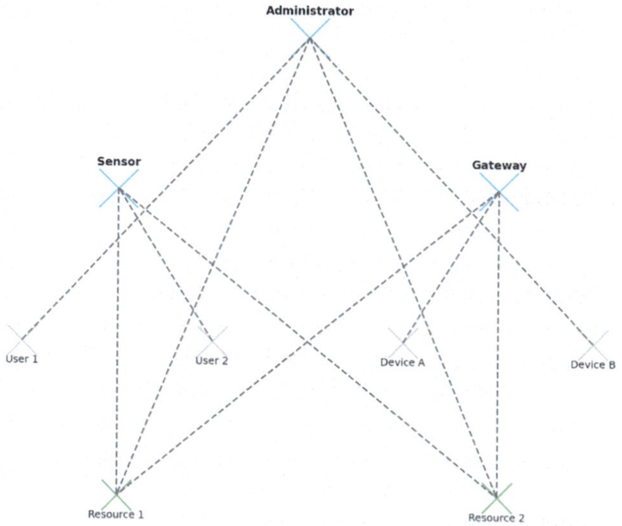

Fig. 3 Role-Based Access Control (RBAC) model example

see Fig. 3) are enforced, often by the IdP or a linked service. For instance, an IdP may check that a given device identity is allowed to read a sensor or execute an actuation command before granting access.

While centralized identity architectures have the benefit of a well-established trust anchor and mature technology support, they suffer from several limitations that become increasingly problematic in WoT-scale deployments:

- **Single point of failure (see Fig.** 4) **and security risks**: The central IdP is a critical dependency—if this server fails or is compromised, the entire network's

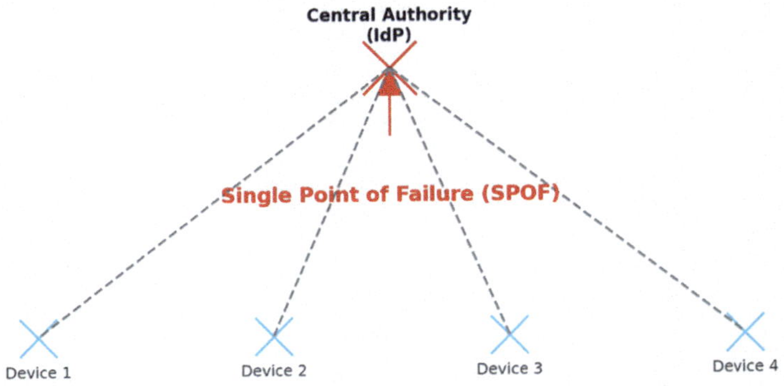

Fig. 4 Centralized architecture where the IdP is a SPOF

identity system collapses. All authentication and authorization services depend on it being available and secure. Attackers also recognize the IdP as a high-value target: a successful attack could expose every device identity and credential in the network. This concentration of trust is risky; a breach can have catastrophic consequences, allowing impersonation of devices or unauthorized network access. Even without a breach, an overwhelmed or crashed IdP can cut off legitimate device communication. In a large-scale WoT scenario with millions of devices, this single point of failure is a serious threat to reliability and security.

- **Scalability bottlenecks**: As the number of connected devices grows exponentially, a centralized identity server can struggle to keep up with authentication requests and data management. The IdP must provision storage for vast amounts of identity data and handle frequent verification queries from devices. This can lead to performance degradation—higher latency in authentication, throughput limits, and potential downtime under peak loads. In the WoT, where devices may be constantly connecting and exchanging credentials (sometimes in real time or in bulk), a central identity service can become a traffic bottleneck. Over time, scaling such infrastructure (adding more servers, load balancers, etc.) becomes complex and costly and still retains the fundamental central bottleneck design.
- **Limited autonomy and vendor lock-in**: In centralized systems, device owners and even devices themselves have little control over identity data or how authentication occurs. All trust decisions are mediated by the central authority's policies. This can lead to vendor lock-in—if the identity management system is part of a proprietary platform (e.g., a cloud IoT service), devices are tied to that platform's rules and interfaces. Switching to a different system or customizing identity workflows is difficult. For example, a company might be forced to use a cloud provider's identity service for its sensors, limiting integration with sensors managed by another provider. This rigidity is ill-suited to the dynamic and diverse WoT environment, where devices from different vendors and domains need to interoperate. A single centralized scheme cannot easily accommodate the heterogeneity of WoT devices and usage scenarios, often lacking flexibility for local or user-defined identity policies.

- **Privacy concerns**: Centralized identity management accumulates sensitive information in one place. The IdP and its identity database may hold not only device credentials but also potentially personal data and metadata about device usage. This aggregation poses privacy risks. Unauthorized access to the central repository can leak data about all devices. Even legitimate IdP operators might exploit the data—for instance, mining the identity logs [8] or attributes for analytics or monetization, possibly without user consent. Additionally, if the IdP shares data with third parties or is subpoenaed, device owners might have no say in how their devices' identity information is disclosed. These issues erode trust: users and organizations may be uncomfortable connecting devices to a system where so much sensitive information is centrally stored and potentially accessible beyond their control. The need for privacy-preserving or decentralized solutions arises in response to these concerns.
- **Compliance and regulatory challenges**: IoT deployments today span across countries and regulatory regimes, meaning an identity management system may have to adhere to laws like the European GDPR [9] or California's CCPA [10] when handling personal or sensitive data. A centralized identity service becomes a choke point for compliance—it must implement all relevant data protection rules correctly for each jurisdiction. If a single IdP serves a global WoT, it faces the burden of multi-jurisdictional compliance, which is complex and expensive to manage. Any lapse could result in legal penalties or shutdown of device services in certain regions. Decentralized or federated approaches can potentially ease this by localizing identity data, but traditional designs typically did not consider such fragmentation, leaving a gap between centralized identity management and regulatory requirements.

In summary, while centralized identity management has been the default for many early IoT systems due to its simplicity and reuse of existing enterprise solutions, it falls short in the context of a massive, diverse, and longevity-focused Web of Things. The SPOF [11] vulnerability, scaling issues, lack of user control, privacy risks, and compliance difficulties all point to the need for rethinking identity management. Decentralization offers a pathway to address these limitations by distributing trust and control, as we explore next [12].

3 Blockchain and Decentralized Identity Management

3.1 Introduction to Blockchain for Identity Management

Blockchain technology introduces a paradigm shift in how trust can be established in distributed systems. At its core, a blockchain is a decentralized ledger maintained by a network of participants rather than a single authority. Several key features make blockchain an attractive foundation for WoT identity management:

- **Decentralization**: Because the ledger is shared across many nodes, no single party exclusively controls the data. This can eliminate the single point of failure inherent in centralized IdPs—even if some nodes go offline or malicious, the system as a whole can continue to operate correctly. For WoT, this means device identity information could be resiliently stored and verified without depending on any one server or company.
- **Immutability**: Entries in the blockchain are tamper-resistant; once a transaction is confirmed and added to the chain, it cannot be easily altered or deleted. This immutability provides a trusted audit trail of identity events. For example, a device's identity registration, credential issuances, or revocation could be permanently recorded, making it easy to detect unauthorized changes or identity forgeries. The immutable ledger thus helps ensure the integrity of identity data in WoT.
- **Security through cryptography**: Blockchain systems employ strong cryptographic mechanisms (hashing, digital signatures, public/private key cryptography) to secure transactions. Identities and credentials anchored on a blockchain benefit from these properties—e.g., a device identity could be represented by a public key hashed into the ledger [13], and only the device with the corresponding private key can prove control of that identity. The distributed consensus algorithms make it extremely difficult for an attacker to tamper with identity records or create fraudulent identities without controlling a majority of the network's power. This level of security is particularly useful for WoT, where devices might be deployed in untrusted environments.
- **Transparency and verifiability**: All participants have access to the ledger's contents (in public or consortium blockchains), which means identity assertions can be independently verified by any stakeholder at any time. A service wanting to verify a device's credentials doesn't need to contact a central IdP; it can directly check the blockchain's state. This trust by verification reduces the need to *trust intermediaries* and instead allows parties to trust the blockchain protocol. In WoT, it enables *peer-to-peer trust*: devices and services can validate each other's identities through the ledger.

By virtue of these features, blockchain-based identity management addresses many challenges of the traditional model. Decentralization and consensus remove the single trusted authority bottleneck, greatly enhancing fault tolerance and making it harder for attackers to compromise the system. The scalability can improve as well: rather than funneling all requests to one IdP, identity validations can be distributed across the blockchain network. Although public blockchains have throughput limits, new consensus mechanisms and off-chain techniques can boost performance (discussed later in Sect. 5).

It's important to note that using blockchain does not mean making all device data public; rather, blockchain would store references or proofs of identity information, while sensitive details can remain encrypted or off-chain. In practice, a WoT identity system might record a device's decentralized identifier or a hash of its credentials on-chain, and use blockchain transactions to manage updates or revocations, all without exposing raw private data. In doing so, blockchain acts as a decentralized trust anchor—all WoT participants can trust the authenticity of identity data on the ledger without needing a central administrator, because the network's consensus guarantees its validity. Overall, blockchain technology offers a compelling alternative that can eliminate many limitations of traditional solutions. By distributing trust and using an immutable ledger, it aligns well with the WoT's needs for scalability, security, and cross-domain trust. Next, we examine the core components that make up a blockchain-based decentralized identity management approach, particularly how decentralized identifiers, self-sovereign identity principles, and smart contracts work together to enable device identity management in the WoT.

3.2 *Core Components of Decentralized Identity*

Decentralized Identifiers (DIDs). A DID is a new type of identifier defined by the W3C that enables verifiable, decentralized digital identity. Unlike traditional identifiers (URLs, usernames, etc.) which are issued by central authorities or tied to specific services, a DID is created by the identity owner (in this case, a device or its owner) and does not require a central registry. A DID typically looks like a URI (e.g., did:example: 123456) and is associated with a DID document containing public keys and service endpoints for that identity. In a WoT setting, when a device is onboarded, it can be assigned a unique DID rather than being registered in a central database. This DID and its initial public key are written to the blockchain (e.g., via a registration smart contract), establishing an immutable, globally verifiable record of the device's identity. Because the blockchain is decentralized, the DID is not owned by any single provider—the device owner can control it directly by managing the associated cryptographic keys. The role of DIDs in device identification is to provide a universal handle for the device that any other party can resolve (using the blockchain or a DID resolution protocol) to obtain the device's public key and other metadata needed to interact with it securely. This eliminates reliance on a centralized lookup service. For example, if two devices from different manufacturers need to interact, each can trust the other's DID record on the blockchain as the source of truth for that device's identity, rather than having to trust each other's manufacturer or a third-party directory.

3.2.1 Self-Sovereign Identity (SSI)

SSI is an approach to digital identity that emphasizes user (or entity) control and privacy. Applying SSI principles to WoT means giving device owners sovereignty over their device identities and credentials, rather than outsourcing that control to a vendor or authority. In practice, under SSI each device (or its owner) maintains a wallet or agent that stores its *Verifiable Credentials (VCs)*—tamper-evident digital credentials issued by various authorities about the device. For instance, a thermostat might have a VC from its manufacturer stating its model and a VC from a regulatory body certifying compliance with safety standards. Under an SSI model, these credentials are stored with the device/owner and are presented when needed, rather than being siloed in a central server. The benefits for device owners are significant: they can choose which credentials to share and with whom, and they *only need to disclose the minimum information necessary*. For example, to join a smart home network, a device might prove "I am a certified thermostat from Manufacturer X" without revealing its serial number or any data beyond what's required to establish trust. This is often achieved with selective disclosure techniques or zero-knowledge proofs. The blockchain complements SSI by acting as a public ledger for verification—issuers can anchor hashes of credentials on-chain, and devices can publish their DIDs, so when a device presents a credential, the verifier can check the blockchain to ensure the credential is valid (e.g., not revoked and issued by a trusted authority). In our WoT framework, SSI means each device holds its own credentials (in an on-device secure storage or an owner's app) and uses them to authenticate in a peer-to-peer manner, rather than the credentials living in a central IdP database. This self-sovereign model enhances privacy and autonomy: devices prove things about themselves without constant reference to a central authority, and owners have greater control over device data. Additionally, SSI allows portability of identity—if a device moves from one network environment to another, it carries its identity and credentials with it, which can be verified on the blockchain, rather than needing a new identity issued by the new environment.

3.2.2 Smart Contracts

Smart contracts are self-executing code running on the blockchain that can enforce rules and automate processes based on predefined conditions. In the context of decentralized identity management for WoT, smart contracts serve as the logic that glues everything together on-chain. They can automate many identity management tasks that would traditionally be handled by central servers or manual administration. Key uses of smart contracts in our framework include:

- **Automated device onboarding**: When a new device wants to join the WoT, a *registration smart contract* can handle the onboarding. It can verify that the device provides a valid initial proof and then record the device's DID and public key on the blockchain. This process is trust-minimized: the smart contract only allows registration if cryptographic checks pass, ensuring rogue devices cannot falsely register as someone else. The entire onboarding can happen without a human operator, making scaling to thousands of devices feasible.
- **Dynamic access control**: Smart contracts can encode access control policies that automatically grant or deny device actions based on their credentials and roles. For instance, an *access policy contract* might specify that "a device with a valid *TemperatureSensor* credential may send temperature data to the cloud service." When a device attempts an action, the contract can check the blockchain for the

device's DID and credentials (VCs) and execute the policy logic to decide if the action is permitted. This replaces or augments centralized authorization servers. Because the rules are on-chain, they are transparent and enforceable without a central authority's intervention, reducing administrative overhead and chances for error. Policies can be updated through on-chain governance or multi-signature approvals to adapt to new conditions.

- **Secure peer-to-peer interactions**: In WoT, devices often need to interact directly. Smart contracts can facilitate these peer-to-peer exchanges by acting as an impartial broker of trust. For example, before a sensor writes data to an actuator, a contract can require both to present a valid identity (DID) and any required credentials, using the blockchain's records to verify each one. Only if the conditions are satisfied (both identities recognized, credentials check out, and perhaps a usage policy allows it) will the contract log the interaction or allow it to proceed. This ensures that even without a central server, devices don't blindly trust each other—they trust the contract and blockchain verification. Essentially, the smart contract can serve a similar role to a security kernel, but in a decentralized fashion.
- **Lifecycle management**: Smart contracts also handle ongoing identity management tasks like revocation of credentials or identities, updates to keys, and expiration of permissions. For instance, if a device is decommissioned or compromised, an authorized party (such as the device owner or manufacturer) can call a revocation function on a smart contract to mark that device's DID or specific credentials as revoked. The contract would update the blockchain state to reflect this, and from that point on, any attempt by the revoked DID to authenticate or any presentation of a revoked credential would be automatically rejected by the contracts. Similarly, contracts can enforce credential expiration dates and require periodic renewal, or manage key rotations by updating the public key associated with a DID when a device's keys change. All these processes happen transparently and consistently across the network.

By leveraging smart contracts for identity management, the system achieves a high degree of automation and trustworthiness. Rules are executed exactly as written, without needing intermediaries, which reduces both administrative burden and opportunities for malicious interference. Smart contracts essentially embed the security policies into the fabric of the WoT blockchain network. Together, DIDs, SSI, and smart contracts form the foundation of our decentralized identity framework: DIDs provide the *identifiers*, SSI provides the *credentials* and *principles of user control*, and smart contracts provide the *logic* to use those identifiers and credentials securely.

4 Proposed Framework for Decentralized Identity Management

4.1 Framework Overview

Building on the concepts above, we propose a comprehensive decentralized identity management framework for the Web of Things. The framework is designed to eliminate the single points of failure and limitations of centralized approaches by using blockchain as a distributed trust backbone and to empower device owners with greater control (in line with SSI principles) while maintaining strong security. At a high level, the framework integrates the following elements (see Fig. 5):

- **Blockchain-based identity registry**: A blockchain ledger stores essential identity records for devices. This includes device DIDs and pointers to current public keys, as well as hashes or identifiers for verifiable credentials. The blockchain acts as the source of truth for device identities and credential status.
- **Decentralized identifiers and credentials**: Each device is assigned a DID and manages its own credentials off-chain (in a secure wallet or storage). Issuers (like manufacturers or authorities) register credential proofs on-chain, linking them to the device's DID. This way, anyone can verify a credential by comparing the device's presented proof to the blockchain record.
- **Smart contract modules**: A set of smart contracts implements the identity management processes: one contract might handle new device registration, another manages credential issuance and revocation logs, and another enforces access control. These contracts collectively ensure that only legitimate, verified devices can join and act within the WoT environment, without a central administrator, by automatically checking identities and credentials on-chain.
- **User/administrator interfaces**: Tools for device owners or system admins to interact with the identity system—for example, a web dashboard or mobile app to register a new device which under the hood calls the blockchain registration

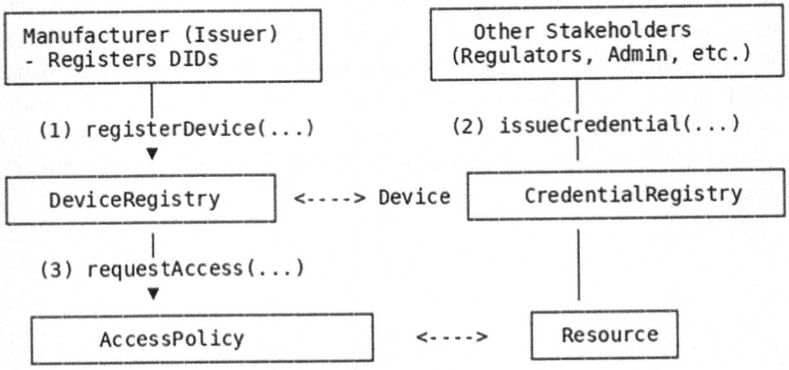

Fig. 5 Blockchain-based identity management framework for WoT

contract, or to revoke a device's identity if needed. These interfaces hide the blockchain complexity and provide a user-friendly way to manage identities in the WoT.

The objectives of the framework are to provide real-time device identity validation, end-to-end security for device interactions, and maximal user/device owner control over identity data. By recording identities and credential attestations on an immutable ledger, the system prevents attackers from impersonating devices or altering identity information unnoticed. By decentralizing the trust, it avoids bottlenecks—multiple network nodes collectively handle identity transactions, and devices can verify each other through the ledger without querying a single server. Furthermore, the framework is extensible: as new devices and credential types emerge, they can be incorporated simply by defining new credential schemas or updating smart contract logic, rather than overhauling a central database.

In summary, our framework aims to build trust into the WoT by integrating blockchain's security with IoT identity needs. It not only secures device authentication and authorization but does so in a way that aligns with the open, interoperable spirit of the Web of Things—identities are not confined to siloed platforms but are part of a decentralized web of trust. The following sections detail the key components and processes of this framework and provide an example scenario illustrating how it works in practice.

4.2 Key Components and Processes

The proposed framework operates through four core steps or phases in managing a device's identity life cycle. These steps, implemented via smart contracts and off-chain interactions, are as follows:

1. **Device registration and DID creation**: When a device is first introduced into the WoT environment (for example, at manufacturing or initial deployment), it is registered onto the blockchain. A unique Decentralized Identifier (DID) is created for the device, if not already assigned by the manufacturer. The device (or an authority on its behalf) invokes a *registration smart contract* to publish the device's DID and associated public key to the blockchain. This establishes an immutable on-chain identity record. From this point, the device can be recognized system-wide by its DID. The registration step often includes verification of an initial proof of identity—e.g., the manufacturer might sign the registration transaction to prove the device is genuine. Once completed, any stakeholder can look up the device's DID on the ledger and obtain its public key and basic identity metadata, which will be used for subsequent trust checks. Above Code 1 represents the solidity pseudo for the DeviceRegistry contract.
2. **Credential issuance:** After basic identity is set up, the device obtains one or more verifiable credentials that assert specific attributes, roles, or qualifications of the device. Credential issuers (which could be the device manufacturer, the owner

Code 1 Pseudo-code for a minimal on-chain device registry

```
contract DeviceRegistry {
    mapping(address => DeviceInfo) devices;

    struct DeviceInfo {
        string did;
        bytes publicKey;
        bool isRevoked;
    }

    function registerDevice(address device, string memory did, bytes
memory pubKey) external {
        // Store DID & publicKey on-chain
        devices[device] = DeviceInfo(did, pubKey, false);
        emit DeviceRegistered(device, did);
    }

    function revokeDevice(address device) external {
        devices[device].isRevoked = true;
        emit DeviceRevoked(device);
    }
}
```

organization, or third-party certifiers) create digital credentials and deliver them to the device (or the device owner's wallet). Each credential is also anchored on-chain for transparency: a cryptographic digest (hash) or a reference of the credential is posted to a *credential registry contract* on the blockchain. This step ensures that anyone can later verify the authenticity and validity of the credential by checking it against the blockchain record. Examples of credentials include: a *capability certificate* stating the device's functions (e.g., "temperature sensor, accuracy ±0.5 °C"), a *ownership credential* linking the device to its owner, or a *compliance certificate* for regulatory adherence. By storing proofs of these credentials on-chain, the framework makes credentials universally discoverable and verifiable without central authority. The device holds the full credentials off-chain and can present them when needed, while verifiers use the on-chain entry to ensure the credential was indeed issued by a trusted party and is not revoked. Above Code 2 represents the solidity pseudo for the CredentialRegistry contract.

3. **Authentication and authorization:** Whenever the device attempts to interact within the WoT—for instance, connecting to a service, sending data to a platform, or requesting to perform an action—the framework invokes an authentication and authorization process mediated by smart contracts. Essentially, the device must prove its identity (via its DID and an associated private key signature) and may need to present relevant credentials. A dedicated *access*

Code 2 Pseudo-code for credential management.

```
contract CredentialRegistry {
    struct Credential {
        address issuer;
        string credentialId;
        string deviceDID;
        bool revoked;
    }

    mapping(address => Credential[]) credentials;

    function issueCredential(
        address device,
        string memory credentialId,
        string memory deviceDID
    ) external {
        / Create credential, store on-chain
        credentials[device].push(Credential(msg.sender,
    credentialId, deviceDID, false));
        emit CredentialIssued(device, credentialId);
    }

    function revokeCredential(address device, string memory
credentialId) external {
        // Mark credential as revoked if msg.sender == issuer
        // ...
        emit CredentialRevoked(device, credentialId);
    }
}
```

control smart contract checks the blockchain to confirm the device's DID exists and is not revoked, then verifies that the device's presented credentials (if any are required for the action) match the records (issuer signatures, non-revocation, etc.). If everything is in order and the device's attributes meet the policy for the requested action, the smart contract grants authorization for the interaction. This could mean allowing the device to publish data to a certain topic, or permitting it to invoke a function on another smart device. The outcome of the check (granted or denied) can also be logged on-chain or off-chain for auditing. This step is repeated as necessary for ongoing interactions, providing continuous verification. Notably, because it's handled by code, this process is both real-time and consistent–every device is judged by the same rules, eliminating bias or human error, and decisions are made in a fraction of a second, suitable for

Code 3 Pseudo-code for device authentication and authorization

```
    contract AccessPolicy {
        DeviceRegistry deviceRegistry;
        CredentialRegistry credentialRegistry;

        function requestAccess(bytes32 resourceId) external view
view returns (bool) {
            // 1. Retrieve device info
            DeviceInfo memory info =
deviceRegistry.devices(msg.sender);
            require(!info.isRevoked, "Device revoked");

            // 2. Check for valid credentials
            Credential[] memory creds =
credentialRegistry.credentials(msg.sender);
            bool validCredentialFound = false;
            for (uint i = 0; i < creds.length; i++) {
                if (!creds[i].revoked &&
keccak256(bytes(creds[i].deviceDID)) ==
keccak256(bytes(info.did))) {
                    validCredentialFound = true;
                    break;
                }
            }
            require(validCredentialFound, "No valid credentials");

            // If checks pass, access is granted
            return true;
        }
    }
```

automated WoT operations. Above Code 3 represents the solidity pseudo for the AccessPolicy contract.

4. **Revocation and lifecycle updates:** Over a device's lifetime, its identity status or credentials may need to change. The framework includes mechanisms for revocation and updates via smart contracts. If a device is decommissioned, lost, or determined to be compromised, an authorized party (such as the owner or an administrator) can call a *revocation function* on the blockchain to invalidate the device's DID or specific credentials. This action updates the on-chain state (for example, setting a flag that the DID is revoked or adding the credential to a revocation list). Thereafter, any attempt by that device to authenticate will be rejected by the smart contracts, as the ledger indicates its identity is no longer trusted. Similarly, credentials can be revoked by their issuers through on-chain

Table 4 Example of an off-chain DID Document referencing service endpoints

```
    // Off-chain representation (simplified)
    {
      "id": "did:example:123456",
      "publicKey": "0xABCDEF...",
      "service": [
        {
          "type": "DeviceServiceEndpoint",
          "serviceEndpoint":
"https://api.manufacturer.com/device/123456"
        }
      ]
    }
```

updates if, say, a certification expires or is rescinded. Besides revocation, this step also handles lifecycle updates such as key rotations (posting a new public key for the DID if the device's keys are changed as a security practice) and credential renewals or additions. All changes are recorded on the blockchain, ensuring that participants always refer to the latest valid state of each device's identity. These controls are crucial for maintaining long-term trust: they provide a way to invalidate compromised devices swiftly and globally and to keep identity data current despite the evolving nature of IoT devices (which might change owners, update firmware, etc., over time). Above Code 4 represents the off-chain component of the DID document.

Through these four phases, the framework covers the full identity management lifecycle (see Fig. 6) in a decentralized manner. It's worth noting that much of the complexity (blockchain transactions, cryptographic verification) is hidden behind the scenes; from a user or administrator's perspective, adding a device or granting it permissions can be as simple as clicking a register button on a dashboard, and revoking a device might be a single command. Under the hood, those actions translate into blockchain transactions and contract calls that enforce security policies consistently.

4.3 Example of Our Framework

To illustrate how the proposed framework operates in practice, consider a smart home scenario where a homeowner adds a new smart thermostat to their WoT ecosystem. This example will walk through the steps from device onboarding to normal operation and eventual revocation, demonstrating the trust-building process:

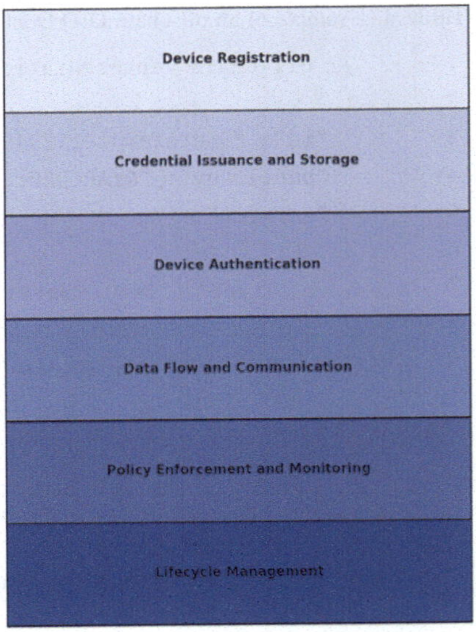

Fig. 6 Device cloud platform workflow

- Onboarding and registration: The homeowner purchases a thermostat from Manufacturer A. Upon installation, they use a smartphone app (provided as part of the smart home system) to onboard the device. The app connects to the thermostat and obtains proof of its origin (the thermostat holds a factory-provisioned private key and certificate from Manufacturer A). The app then calls the device registration smart contract on the blockchain network. It submits the thermostat's public key and a newly generated DID, along with the manufacturer's cryptographic attestation. The smart contract verifies the signature (see Code 5) from Manufacturer A against Manufacturer A's known public key (which is stored on the blockchain as a trusted issuer). Verification passes, so the contract records a new identity entry: DID did:home:thermo1001 with the thermostat's public key, and marks it as active. This transaction is added to the blockchain block. Now the thermostat is officially recognized in the home's WoT blockchain ledger as *Thermostat 1001* with a certain DID. Any component in the home or even external services can query the blockchain for did:home:thermo1001 to get its public key and confirm that it's a registered device.
- key or the home hub's key. A hash of this credential is then written to the blockchain via the credential registry contract, under the thermostat's DID. Additionally, suppose the thermostat needs to prove to a utility company's service that it's an energy-efficient device to integrate with demand-response programs. The utility company (as a third-party issuer) could issue a certificate of energy efficiency as a VC. The thermostat (or homeowner) would register that credential's reference on-chain as well. At this point, the thermostat's identity on the blockchain has pointers to several credentials: manufacturer QA, ownership, and efficiency cert. The device stores the full details of these credentials locally

Table 5 Off-chain signature generation (device) + on-chain verification (application policy)

```
// Off-chain step: Device signs a request with its private key
SignedRequest request = signRequest(deviceDID, deviceVC);

// On-chain step (simplified):
function interactWithService(bytes32 resourceId, SignedRequest
request) external {
    require(verifySignature(request), "Invalid signature");
    require(isDeviceValid(request.deviceDID), "Device
revoked");
    require(hasValidCredential(request.deviceVC), "No valid
credential");
    // ... Additional checks ...
    grantAccess(resourceId, request.deviceDID);
}
```

and will present them when required, while verifiers can check the blockchain to ensure those credentials are valid and signed by recognized authorities.

- **Authentication and normal operation:** Now that the thermostat is set up, it begins normal operation. It periodically needs to send temperature readings to the homeowner's cloud service and sometimes directly interacts with the home's smart HVAC system to adjust heating. Each time the thermostat connects to the cloud or sends a command to the HVAC, the access control smart contract comes into play. When connecting to the cloud, the thermostat's firmware uses its private key to sign an authentication request that includes its DID. The cloud's WoT gateway receives this and invokes a blockchain query or transaction to validate it. The smart contract checks: Is did:home:thermo1001 present and active on the ledger? (Yes.) Is the signature valid for the public key on file? (Yes.) Have there been any revocations? (No.) Thus, the contract (or a blockchain oracle service) confirms the identity, and the thermostat is authenticated. Because the cloud knows this device's DID and has linked it to the homeowner's account through the on-chain ownership credential, it authorizes the device to send temperature data. For a more complex interaction, say the thermostat attempts to tell the HVAC to turn on the heat: the thermostat communicates to the HVAC's controller, which triggers an access control check. The policy (encoded perhaps in another smart contract or in the HVAC controller's trust logic) might require that *only a device with a valid owner credential can command the HVAC*. The thermostat presents a proof of its ownership credential. The HVAC's system verifies this by checking the blockchain: it finds the hash of a valid ownership credential issued to the thermostat's DID by the homeowner and sees no revocation. Satisfied, it allows the command. All of this happens in seconds and without any central server mediating trust—the devices and cloud

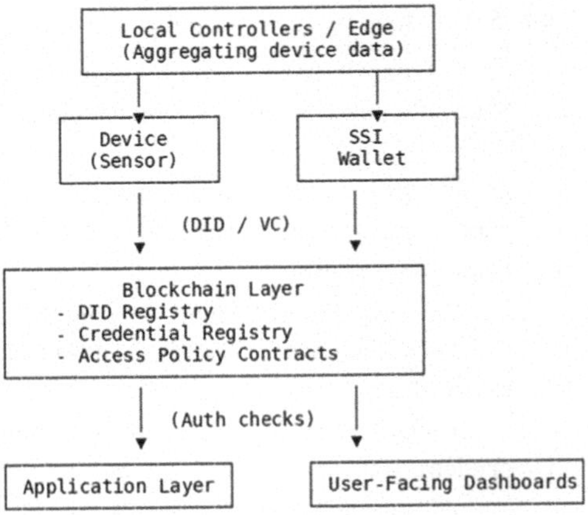

Fig. 7 Layered architecture illustrating DID and SSI integration with on-chain components

components rely on the shared blockchain ledger and smart contracts to validate identities and permissions.

- **Revocation (Compromise and Recovery):** Some time later, suppose the homeowner learns that this model of thermostat has a security flaw that malware can exploit. The manufacturer decides to revoke the identity of any compromised thermostats until they are patched. Using their authority, Manufacturer A issues a revocation transaction on the blockchain for the affected DIDs. In our case, did:home:thermo1001 is marked as revoked in the device registry (see Fig. 7) via the smart contract. Immediately, this change propagates to all participants: the next time the thermostat tries to authenticate or send a command, the access control smart contract will see the "revoked" status and deny the request. The homeowner's cloud service would refuse data uploads from the thermostat, and the HVAC would ignore its commands. In effect, the thermostat is quarantined from the WoT environment for safety. The homeowner is notified and updates the device's firmware. After resolving the issue, the homeowner (or manufacturer) can *restore* the device's identity by updating its status on-chain (or issuing a new DID and credentials if needed). The blockchain now shows the device as active again, possibly with an updated public key if the compromise involved key exposure. Normal operation can resume, with confidence that previously compromised credentials are no longer accepted. This demonstrates how the framework handles the full lifecycle, including failures: trust can be swiftly revoked and later renewed in a transparent, consistent manner.

This scenario highlights the benefits of the decentralized approach. The homeowner and devices could operate with minimal trust in external parties—they trust the blockchain network and issuers they choose (manufacturer, utility), but they did not rely on any single company to broker all identities. If Manufacturer A's servers were offline, it wouldn't matter after initial setup; the ledger carries the needed info. If the homeowner changed to a different home automation provider, their thermostat's identity could port over since it's anchored in a neutral blockchain.

Security incidents are easier to contain through global revocation, and privacy is enhanced (the thermostat only shares what it needs to, and much of the verification uses on-chain hashes rather than exposing credential details). This example thus demonstrates how trust is built and maintained in the WoT using decentralized identity through secure initial registration, credentialing, continuous verification via smart contracts, and robust processes for update and revocation.

5 Challenges and Solutions

Adopting a blockchain-based decentralized identity framework in the Web of Things is not without its difficulties. Several challenges must be addressed to make such a system practical, scalable, and secure in real-world deployments. We discuss some of the key challenges below, along with potential solutions and mitigations for each:

- **Scalability:** While decentralization removes the bottleneck of a central IdP, blockchain platforms face their own scalability limits. Public blockchains like Ethereum can handle only a limited number of transactions per second, and as WoT devices might generate a high volume of identity transactions (registrations, credential updates, authentication checks), the network could become congested. This can lead to high latency or transaction fees, undermining the benefits. The solution is employing Layer-2 scaling solutions and optimized consensus mechanisms to increase throughput (see Fig. 8). Techniques such as payment channels/state channels (where devices perform frequent interactions off-chain and only commit final states to the blockchain) and sidechains or rollups (which bundle many identity operations into a single batch) can dramatically reduce

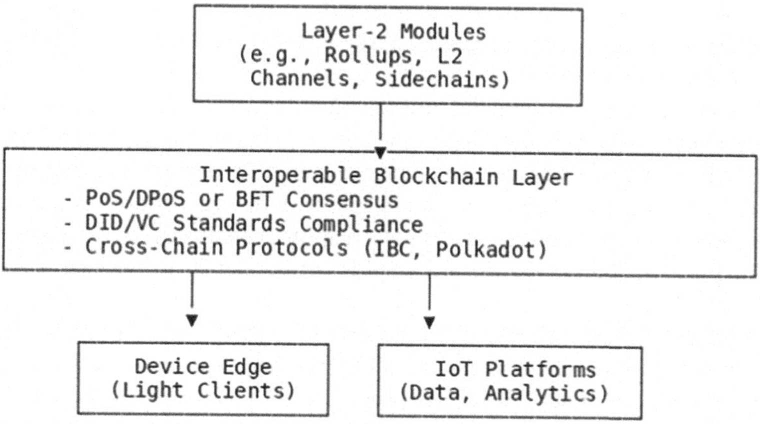

Fig. 8 Layer-2 scaling approaches (top) and cross-chain communication (middle).

the load on the main chain. For example, 1000 authentication events could be verified off-chain and only one aggregated proof posted on-chain. Additionally, permissioned or consortium blockchains specifically set up for WoT identity might use more efficient consensus (like Proof-of-Authority or raft algorithms) to handle device volumes. These approaches have shown potential to increase transaction throughput by orders of magnitude, making decentralized identity feasible even as billions of devices come online. In resource-constrained IoT contexts, the combination of on-chain and off-chain processes ensures scalability without requiring every tiny device to directly interact with a global ledger for every action.

- **Interoperability:** The WoT spans many platforms, industries, and even different blockchain networks. A challenge arises in ensuring that a device's decentralized identity is recognized across disparate systems. For instance, a device might have a DID on one blockchain, but a service it wants to use verifies identities on another network. Similarly, standardizing how identity data is exchanged is difficult when multiple protocols and data formats exist. *Solution:* Develop and adopt universal standards and cross-chain interoperability protocols for identity. On the standards front, the W3C DID specification and verifiable credentials data model provide a baseline that all parties can implement, ensuring that a credential issued in one context can be understood in another. For cross-chain scenarios, emerging technologies like Inter-Blockchain Communication (IBC) (from the Cosmos project) and Polkadot's cross-chain messaging allow different blockchains to talk to each other. In practice, this could mean a WoT identity blockchain could trust and accept credentials anchored in another chain, or seamlessly transfer a device's identity record if it moves between domains. Another aspect of interoperability is semantic: devices and services need to agree on the meaning of identity attributes. Efforts to create common ontologies and schemas for device credentials (so that "temperatureSensor" or "batteryLevel" have consistent definitions, for example) are crucial. By leveraging these solutions, the decentralized identity framework can become interoperable across heterogeneous ecosystems, enabling a true *web* of things where trust follows the device wherever it goes.
- **Privacy:** One might assume that placing identity information on a public ledger could compromise privacy, since blockchains are often transparent. Indeed, a poorly designed identity system could inadvertently expose sensitive metadata (like which devices belong to which user, or usage patterns gleaned from credential updates). *Challenge:* How to reap the benefits of blockchain's transparency and auditability without divulging private details of devices or owners? *Solution:* Use privacy-preserving techniques such as *selective disclosure* and zero-knowledge proofs (ZKPs) to minimize on-chain data. In our framework, we already limit on-chain information to DIDs and cryptographic hashes rather than raw personal data. To further protect privacy, devices should only reveal necessary information: for example, a device can prove it has a valid credential (using a ZKP) without revealing the credential's contents or even which specific credential it is, aside from the fact that it's trusted. Zero-knowledge proof

systems and protocols like zk-SNARKs or Bulletproofs could allow a smart contract to verify that "device X possesses a credential signed by authority Y with attribute Z" is true, without the device posting that credential openly. Another approach is using permissioned blockchains or shard chains where identity data is confined to a subset of nodes that are authorized, combined with *cryptographic access control* so that only those with a need-to-know can read certain fields. Additionally, rotating identifiers and pseudonymous DIDs can prevent correlation of activities to a single device over time. In essence, careful architecture can ensure the system is transparent for verification but opaque for sensitive information, thus preserving privacy while still enabling trust. Techniques like did:peer (pairwise DIDs) could be used for devices to have different identifiers with different services, preventing cross-context correlation.

- **Regulatory compliance:** Decentralized systems challenge traditional regulatory models. Regulations often assume a clear data controller who can ensure compliance (for example, deleting user data upon request in compliance with GDPR's right to erasure). In a blockchain-based identity system, data is immutable and there is no single controller, which raises questions about how to fulfill such obligations. *Challenge:* How to design the system so that it respects laws and regulations regarding data protection, identity, and cybersecurity? *Solution:* Architectural and governance measures can bridge this gap. One approach is to use off-chain storage for personal data linked to on-chain identities, where that off-chain store can be designed to allow deletions or modifications as required by law (the on-chain ledger would just store non-sensitive pointers or hashes). This way, the blockchain avoids holding raw personal information that might conflict with regulations. Another measure is to implement a form of consortium governance for the identity blockchain: participants (companies, authorities) agree on policies that align with regulations, and smart contracts can be coded to enforce certain compliance rules (for example, automatically expiring certain data after a time to satisfy data minimization principles). Permissioned blockchains can confine data within a region or set of trusted operators to address data locality requirements (e.g., EU data stays in EU nodes). Finally, close collaboration with regulators to create sandboxes and get clarity on decentralized identity practices will help. The framework should be flexible enough to incorporate "regulated nodes"—perhaps certain organizations act as privacy guardians or auditors on the network to ensure it meets legal standards. By proactively including compliance as a design goal, and possibly allowing emergency intervention mechanisms (like a legal order could trigger a contract to freeze an identity if absolutely required), the system can maintain decentralized ideals while avoiding legal roadblocks.

In summary, while decentralized WoT identity management promises improved scalability, security, and user empowerment, it must overcome scalability constraints, interoperability hurdles, privacy pitfalls, and regulatory obstacles to be viable at global scale. Fortunately, ongoing research and development in the blockchain and identity communities are actively addressing these issues. Layer-

2 scalability, cross-chain standards, privacy-enhancing cryptography, and hybrid on/off-chain models provide a toolkit of solutions. By incorporating these into the framework, we move closer to a practical, large-scale decentralized identity system for the Web of Things.

6 Conclusion and Future Work

The rapid growth and diversity of the Web of Things have exposed the limitations of traditional, centralized identity management. As billions of devices—from consumer wearables to industrial controllers—come online and interact, centralized models struggle with scalability, security, privacy, and interoperability challenges. In contrast, this chapter has shown that blockchain-based decentralized identity management is a robust alternative that can meet the demands of large-scale WoT ecosystems. By leveraging blockchain's immutable ledger and distributed trust model, combined with self-sovereign identity principles and decentralized identifiers, we can fundamentally enhance how device identities are managed and secured. Our proposed framework demonstrates how these technologies can be integrated to provide a *secure, scalable, and user-centric* identity solution for WoT. Through blockchain smart contracts, devices are able to self-register and authenticate without relying on a single authority, eliminating single points of failure and failure-prone centralized oversight. The use of DIDs and SSI means each device (and its owner) gains greater control over identity data—devices hold their own credentials and reveal only the minimum necessary information for a given interaction, enhancing privacy and autonomy. Meanwhile, trust is maintained via the blockchain: devices trust the ledger for identity verification, and the ledger's consensus ensures no malicious actor can undetectably subvert identities. We illustrated in a smart home scenario how the framework enables seamless onboarding, fine-grained access control via verifiable credentials, and swift revocation in case of compromise, all through decentralized mechanisms. Importantly, this approach does not sacrifice security—it uses strong cryptography and the collective enforcement of the network to actually *increase* security relative to the siloed, opaque operation of many IoT systems today. It also lays a foundation for interoperability: identities anchored on a blockchain can be universally recognized across different platforms, fostering a more open WoT ecosystem.

Despite these advancements, our exploration also highlighted that significant future work remains to fully realize the vision of decentralized identity in WoT. Key areas for continued development include:

- **Scalability enhancements**: Further research into efficient on-chain/off-chain hybrid models, next-generation consensus algorithms, and IoT-friendly blockchains is needed to ensure that identity management can keep pace with potentially *billions* of transactions and devices. The use of advanced layer-2 solutions and optimized crypto protocols (like aggregating many signatures into one) will be critical.

- **Interoperability and standards:** The community must converge on standards for representing and exchanging device credentials and policies. Efforts to integrate different blockchain networks (through cross-chain identity bridges or federations) will be crucial for a truly global solution. This also involves creating mappings between WoT standards and identity standards so that devices can move between contexts smoothly.
- **Privacy-preserving techniques:** Applying cutting-edge cryptography, such as zero-knowledge proof systems, in resource-constrained IoT devices is an open challenge. Techniques that are computationally light yet powerful enough to protect privacy in decentralized identity exchanges will be a focus. Also, UI and UX research is needed to make privacy choices understandable to users who manage device identities.
- **Regulatory and governance models**: As the technology outpaces policy, interdisciplinary work is needed to align decentralized identity with legal frameworks. Developing governance models for identity blockchains (who runs the nodes, who has authority to issue or revoke credentials in certain domains, how disputes are resolved) will be important for adoption in enterprise and public sector applications. Engaging with regulators to update or clarify laws around decentralized data (for example, how to handle GDPR's requirements) is part of this effort.

In conclusion, decentralized identity management using blockchain, DIDs, and SSI offers a promising pathway to address the inherent vulnerabilities of centralized IoT identity systems. By distributing trust, improving scalability through decentralization, and placing users and device owners at the center of control, this approach can significantly strengthen security and trust in the Web of Things. The framework and discussions presented in this chapter serve as a blueprint and motivation for building next-generation WoT systems that are resilient, privacy-preserving, and user-empowering. As standards evolve and the technology matures, we expect to see increasing real-world implementations of these concepts—from smart homes and industries to smart cities—thereby gradually transforming the WoT into a more secure and equitable ecosystem where trust is inherent and decentralized. The journey is ongoing, and the collaboration of researchers, industry practitioners, and policymakers will be key in overcoming remaining challenges and driving widespread adoption of decentralized identity management in the Web of Things.

References

1. A. Gyrard, S. K. Datta, and C. Bonnet, "A survey and analysis of ontology-based software tools for semantic interoperability in IoT and WoT landscapes," in 2018 IEEE 4th World Forum on Internet of Things (WF-IoT), Singapore: IEEE, Feb. 2018, pp. 86–91. doi: https://doi.org/10.1109/WF-IoT.2018.8355091.

2. P. Mahalle, S. Babar, N. R. Prasad, and R. Prasad, "Identity Management Framework towards Internet of Things (IoT): Roadmap and Key Challenges," in Recent Trends in Network Security and Applications, vol. 89, N. Meghanathan, S. Boumerdassi, N. Chaki, and D. Nagamalai, Eds., in Communications in Computer and Information Science, vol. 89. , Berlin, Heidelberg: Springer Berlin Heidelberg, 2010, pp. 430–439. doi: 10.1007/978-3-642-14478-3_43.
3. K. Saito, "WOT for WAT: Spinning the Web of Trust for Peer-to-Peer Barter Relationships," IEICE Transactions on Communications, vol. E88-B, no. 4, pp. 1503–1510, Apr. 2005, doi: https://doi.org/10.1093/ietcom/e88-b.4.1503.
4. D. Pöhn and W. Hommel, "An overview of limitations and approaches in identity management," in Proceedings of the 15th International Conference on Availability, Reliability and Security, Virtual Event Ireland: ACM, Aug. 2020, pp. 1–10. doi: https://doi.org/10.1145/3407023.3407026.
5. E. Barka, S. S. Mathew, and Y. Atif, "Securing the Web of Things with Role-Based Access Control," in Codes, Cryptology, and Information Security, vol. 9084, S. El Hajji, A. Nitaj, C. Carlet, and E. M. Souidi, Eds., in Lecture Notes in Computer Science, vol. 9084. , Cham: Springer International Publishing, 2015, pp. 14–26. doi: 10.1007/978-3-319-18681-8_2.
6. M. C. Domenech, A. Boukerche, and M. S. Wangham, "An Authentication and Authorization Infrastructure for the Web of Things," in Proceedings of the 12th ACM Symposium on QoS and Security for Wireless and Mobile Networks, Malta Malta: ACM, Nov. 2016, pp. 39–46. doi: https://doi.org/10.1145/2988272.2988274.
7. T. S. Sobh, "Identity management using SAML for mobile clients and Internet of Things," JHS, vol. 25, no. 1, pp. 101–126, Feb. 2019, doi: https://doi.org/10.3233/JHS-190606.
8. C.-W. Tsai, C.-F. Lai, M.-C. Chiang, and L. T. Yang, "Data Mining for Internet of Things: A Survey," IEEE Commun. Surv. Tutorials, vol. 16, no. 1, pp. 77–97, 2014, doi: https://doi.org/10.1109/SURV.2013.103013.00206.
9. A. D. Kounoudes and G. M. Kapitsaki, "A mapping of IoT user-centric privacy preserving approaches to the GDPR," Internet of Things, vol. 11, p. 100179, Sep. 2020, doi: https://doi.org/10.1016/j.iot.2020.100179.
10. K. Bellamy, "The Ethical and Legal Implications of IoT Data in Business Organizations," 2024, doi: https://doi.org/10.25776/DXTR-VJ53.
11. S. El Jaouhari, A. Bouabdallah, and A. A. Corici, "SDN-based security management of multiple WoT Smart Spaces," J Ambient Intell Human Comput, vol. 12, no. 10, pp. 9081–9096, Oct. 2021, doi: https://doi.org/10.1007/s12652-020-02601-y.
12. O. Dib and K. Toumi, "Decentralized Identity Systems: Architecture, Challenges, Solutions and Future Directions," AETiC, vol. 4, no. 5, pp. 19–40, Dec. 2020, doi: https://doi.org/10.33166/AETiC.2020.05.002.
13. M. Toorani and C. Gehrmann, "A decentralized dynamic PKI based on blockchain," in Proceedings of the 36th Annual ACM Symposium on Applied Computing, Virtual Event Republic of Korea: ACM, Mar. 2021, pp. 1646–1655. doi: https://doi.org/10.1145/3412841.3442038.

Index

S. El Jaouhari, N. El Madhoun (eds.), *Blockchain-based Web of Things Security: Advances, Challenges, and Solutions*, Signals and Communication Technology,
https://doi.org/10.1007/978-3-032-07671-7

The manufacturer's authorised representative in the EU is Springer Nature Customer Service Centre GmbH, Europaplatz 3, 69115 Heidelberg, Germany. If you have any concerns regarding our products, please contact ProductSafety@springernature.com

Printed and bound by CPI Group (UK) Ltd, Croydon, CR0 4YY
05/10/2026
02228059-0003